A Case Study on Intensification for Freshwater Fish Culture and Training

A Case Study on Intensification for Freshwater Fish Culture and Training

N.K. Bose
Shambhu Dhir

CYBER TECH PUBLICATIONS
4264/3, Ansari Road, Daryaganj, New Delhi-110002 (India)
Ph.: 011-23244078, 011-43559448 Fax: 011-23280028
E-Mail: cyberpublicationsdelhi@yahoo.com
Website: www.cybertechpublications.com

A CASE STUDY ON INTENSIFICATION FOR FRESHWATER FISH CULTURE AND TRAINING

N.K. Bose, Shambhu Dhir

First Edition 2011

Published by :

G.S. Rawat for **Cyber Tech Publications**

4264/3, Ansari Road, Daryaganj, New Delhi-110002 (India)

Ph.: 011-23244078, 011-43559448 Fax: 011-23280028

E-Mail: cyberpublicationsdelhi@yahoo.com

Website: www.cybertechpublications.com

CONTENTS

Preface

The pond culture of Indian, Chinese and the common carps is probably the most ancient system of aquaculture in Asia, although its importance and contribution to fish production differ considerably among the various countries of the region. There are plans for the expansion of carp culture in almost all the countries through the transfer of existing technologies and their adaptation and improvement through research.

2.1 Cultivated Species and Species Combinations

The traditional form of carp culture in South Asia is based on the so-called Indian major carps, catla (Catla catla), rohu (Labeo rohita), mrigal (Cirrhina mrigala), and calbas (Labeo calbasu). The three varieties of common carp (Cyprinus carpio) were introduced into India almost a century ago for culture in certain areas, but the interest in the culture of exotic carps became substantial only some two decades ago with the introduction of the Chinese carps (grass carp, Ctenopharyngodon idella, and silver carp, Hypophthalmichthyus molitrix). As the Indian major carps, like the Chinese carps, do not breed ir stagnant ponds and generally spawn only in their natura habitat, the farmers had to obtain the seed in the form of egg hatchlings, and fry from rivers. The seed of all the specie occurred together in nature and could not be sorted out easi according to species. The farmers, therefore, collected a cultured them together in ponds. Thus a system of polycult evolved.

The zonation in their feeding habits; catla feeding lar on the surface and the mrigal and calbasu at the bottom latter feeding also on snails) gave a scientific basis for combination of species. The CIFRI has conducted conside research on the species ratios and stocking densities tha high yields under a series of culture conditions. In recent years this species combination has been enlarged to include the Chinese and common carps, generally omitting the calbas. In experimental and demonstrations ponds, CIFRI has obtained a rate of production of up to 10 000 kg/ha per annum with fertilization and supple-mental feeding. However, the wide-scale adoption of this practice by farmers would obviously depend on the availability of fry or fingerlings of the various species at the appropriate time and in sufficient numbers. The complexity of culture is also obviously increased when a larger number of species with differing habits and requirements have to be cultured together.

Although the benefits of polyculture are widely accepted and high production has been obtained, it has not yet been established that the combination of species now adopted is the most appropriate one, particularly in view of the overlap of feeding habits of some of the species. It has also not been experimentally proved that monoculture will not yield high and economically acceptable production, rendering the overall culture operations simpler. On the other hand, in areas where the ponds are infested by a large snail population, the addition of a snail feeding fish like the pangas (Pangasius spp.), which has been shown to be a good culturable species, may prove valuable. This could replace the slow-growing calbas, which is now generally omitted from the species combinations.

— *Author*

1

Introduction

BACKGROUND INFORMATION

Research at the Central Inland Fisheries Research Institute (CIFRI) on composite fish culture and implementation of results in field trials all over India have shown that intensive pond fish culture technology can secure very high yields (over 7 000 kg/ha) in this country.

From the worldwide experiences it is known that fish diseases represent a major threat to fish culture in general and especially to intensive biotechnologies. The personal contacts of the consultant with fish culture scientists revealed the presence of an experience or an opinion among some of them that, up to now, fish diseases do not seem to have a significant impact on composite fish culture in India, as managed by methods developed so far. According to these opinions, losses in ponds can be avoided by already known methods in technology, including prevention of predation and of some parasitic invasions of skin and gills. On the other hand, some reports indicate that diseases do occur and have an impact on production in polyculture ponds. These reports as well as the limited laboratory diagnostic work and an experiment conducted at FARTC during the consultancy indicate that pond fish culture in India already stands at the same risks from diseases as the aquaculture in other countries.

Some information is available on fish diseases and their control in India. There are also some data on zoological aspects of fish parasites, on seasonal incidence of some ectoparasites and on gross symptoms of some diseases noted in composite fish culture. Generally, however, there is a lack of well documented research on most etiological groups of diseases and treatments for them. Bacterial and viral diseases were generally not a subject of up to date research by adequate laboratory methodologies. Fish health monitoring is performed only to a limited extent at the field level.

This situation indicates a serious lag in data on diseases and their prevention behind the research and practical results on fish production in ponds. In contrast to the situation in developed countries and in some developing countries, India does not yet have a diagnostic institution or laboratory for fish diseases.

The Project IND/75/031 has therefore included work on fish diseases as one of its main objectives. The FARTC at Dhauli is the first location in India that will have facilities and personnel for research and diagnostic work on all major groups of fish diseases and work on health monitoring in aquaculture. In order to fulfil its future role in the development of this field in India and in countries of this part of world that do not yet have such facilities, the initial planning of work on fish diseases is of utmost importance and can have far reaching consequences. The consultant has therefore included in this report general recommendations on organization of the work and on possible future plans.

SUMMARIZED DESCRIPTION OF WORK

The consultant stayed at FARTC, Dhauli, from 16 October 1980 to 6 January 1981. The stay was prolonged for a dew days upon the request of the project director in order to complete some experiments. A four day break in above stay was made from 27 to 30 December for a visit to Barrackpore on the request of the Director of CIFRI.

There the consultant informed the director about the progress in his work at FARTC. He also went to two research stations of CIFRI to investigate two fish disease problems and to advise the research workers about possible research approaches for solving the etiology of these diseases.

Four young FARTC scientists R.K. Dey, D. Kumar, B.K. Mishra, K. Suresh) were assigned for work in this field. In addition, two CIFRI scientists (R.N. Pal and D.M. Swami) worked with the group for about one month each.

TRAINING

The work of FARTC staff scientists could be started only after adequate training. The programme of training was adjusted to the fact that none of four assigned FARTC scientists had sufficient previous knowledge of fish diseases, fish health monitoring and protection. Special emphasis was placed on topics that are not fully covered in existing English or other literature, and on practical training.

Theoretical Background

The following topics were covered during the training:

- General pathology, main types of pathological processes in fish and their symptoms;
- Characteristics of etiological groups of fish diseases;
- Epizootiology of fish diseases;
- Prevention of diseases in aquaculture;
- Methods of preventive and curative application of drugs to fish;
- Incorporation of health protection measures into pond culture technology;
- Selection, packing and transport of fish samples for examination in the laboratory;
- Diagnostic methods and procedures for examination of fish (gross examination, determining the condition of fish dissection, hematocrit, bacteriology, virology, parasitology).

Practical Training

This was carried out on the following topics:

- Sampling of fish from ponds for examination;
- Laboratory examination of fish for gross lesions by dissection;
- Taking of material from fish for hematological, parasitological, virological and histopathological examinations;
- Microscopic examination of local preparations from external and internal organs for pathological changes, parasites and bacteria;

- Bacteriological examination of fish (preparation of smears, stains and media, inoculation and examination of dishes with media, basic methods for identification of fish pathogenic bacteria, maintenance of isolates);
- Primary cell cultures and cell lines for fish virology (preparation of glassware, media and other materials, preparation of cell cultures and their use for diagnostic work);
- Intraperitoneal administration of drugs to fish;
- Designing and planning of pond experiments on diseases, as well as their prevention and treatment; conducting of experiments and interpreting the results; an experiment was designed, planned, carried out and completed in 12 experimental ponds; a second experiment was started;
- Keeping records on laboratory work and findings.

Organization of the Seminar on Fish Health Protection

Upon completion of the training by the consultant on general aspects of fish diseases, seven scientists gave a seminar for participants in the ongoing course at the KVK and Trainers' Training Centre of CIFRI. The purpose of this seminar was twofold: (i) to give an opportunity to the scientists that were trained by the consultant to confirm and deepen their knowledge, and (ii) to provide trainees at the Centre with basic knowledge in this field and to initiate cooperation in fish health monitoring between field workers and the FARTC researchers.

Other Activities

The consultant gave three lectures on fish diseases and health protection. In all three instances fruitful discussions developed, showing great interest in this field and its importance. Lectures were held for (i) Director and Extension Officers of the Department of Fisheries, Government of Orissa, (ii) staff members and students of the Veterinary School at the Orissa University of Agriculture and Technology, and (iii) scientists of CIFRI at Barrackpore. The lecture under (i) was of special importance for the development of cooperation in work on fish diseases and health protection between FARTC and the Department of Fisheries.

AID IN ESTABLISHING ADEQUATE CONDITIONS FOR RESEARCH

Utilization of Space

Four laboratories and one storage room were allocated by the project director for work on fish diseases and health protection. The first laboratory will serve as the room for dissection and general examination of incoming fish samples, as well as for work on diseases caused by parasites. The second laboratory will accommodate work on histopathology and on bacterial diseases. The third room will be the laboratory for cell culture and virology. The fourth room will serve as the wet laboratory for experiments with fish in aquariums and provide space for washing and sterilization of glassware. Depending on developments in the research and on practical factors, the space may have to be enlarged in the future.

The consultant made recommendations for setting up the sterile chamber for cell culture and virology work, needed adaptations of furniture and suggested plans for setting up the wet laboratory.

Installation of Equipment

Together with the local staff, the consultant installed the equipment that came to FARTC from the Cuttack Station of CIFRI. When this equipment arrived, it was found that it did not function properly and a claim was sent to the supplier.

Other Activities

In order to secure information about availability of commonly used medicines, the consultant requested and obtained information and literature from several pharmaceutical companies in India. Collection of such information will be continued by staff to have on hand broad information about locally available drugs for research and field use and their prices. To facilitate the diagnostic and research work, the consultant prepared up to date flow charts for isolation and identification of fish pathogenic viruses and bacteria. The charts were to be prepared as wall posters.

The consultant also discussed with and advised the FARTC staff and the potential producers about the design of experimental aquaria, troughs and tanks that will be procured by FARTC for research purposes. Cooperation of scientists assigned to work with the consultant and other researchers and research groups was initiated. On one hand, diagnostic information and preventive treatment recommendations were given for experimental fish. On the other hand, advice and cooperation was utilized or planned with economists, nutritionists, water chemists and statisticians.

RESULTS OBTAINED DURING PRACTICAL TRAINING

Laboratory Work

Diagnostic Work

During the consultant's stay in Dhauli, 101 fish were dissected and their organs examined microscopically. Material for bacteriological examination was taken from 28 fish, for parasitological examination from 10 specimens and for histopathological examination from 35 fish. In addition to pathoanatomical diagnoses, tentative bacteriological diagnosis was set up for 20 fish, final histopathological diagnosis in 4 fish and parasitological diagnosis in 82 fish. Materials under processing and isolated strains of bacteria will be used for setting up final diagnoses and for research purposes.

Examined according to species were 66 rohu, 10 catla, 7 mrigal, 3 common carp and 15 silver carp.

Of the above fish, 66 originated from ponds at FARTC, 27 from nearby village ponds, 5 from the Trainers' Training Centre ponds, and 3 from the Cuttack hatchery. Another break-down from the records shows that 76 of these fish were examined directly after netting from ponds. Two of them had malformations, while the others did not show external signs of disease. There were 25 fish brought for examination from laboratory experiments set up by groups of scientists working on nutrition and on genetics or from the pond experiment set by the fish disease group.

The most striking finding was that 100 percent of the rohu examined suffered from severe pathological changes in their kidneys, detectable only by microscopic examination. Changes seemed to be more pronounced and rather severe in two year fish. They are sufocate to chronic. Findings were confirmed by histopathological examination. Fish had necrotic and degenerative changes in most of tubules (nephrosis) and focal accumulation of a yellowfish material, which was sometimes surrounded by a weak chronic connective tissue encapsulation. The yellowfish micronodules were less abundantly present in liver and in one case very abundantly in ovaries. In most of these fishes, myxosporidian specks were detected in kidneys. It is difficult to give an opinion now whether there is a causal or any relationship between the myxosporidian parasite, the nodules and the nephrosis. Since such a 100 percent rate of heavy kidney damage, as well as the occasional damage of other organs, definitely must have an influence on the condition, growth rate and food utilization of the affected rohu, this problem requires thorough further investigation.

The second important finding in the laboratory was that, after netting and handling (material available so far was limited only to fish in experiments), rohu and common carp suffer from symptoms and mortality typical of the columnaris disease. A Flexibacter columnaris-like myxobacterium was isolated in pure culture from kidneys of several affected specimens and was abundantly present in skin lesions. This problem is of known significance in warmwater pond culture of North America and Europe. The disease can affect almost all fishes but there are some variations in species susceptibility. In some species it occurs almost regularly after netting and stocking. It also develops spontaneously in open waters and ponds, often affecting only gills. The findings here indicate that, in all cultivated species, prevention of this disease may have practical significance for reducing losses after netting, handling and stocking operations. Research in this field is urgently needed.

A possibly significant problem was detected by diagnosing enteritis in mrigal and the finding of the intestinal protozoan parasite from the germs opalina, which has not been reported in fishes from temperate zones. Further work on fish health monitoring should indicate the incidence and importance of this condition and the role of this parasite in it.

Presence of yellow clusters of cyst-like units in the intestinal content of some fish, which could eventually be spores of evecidian parasites, also requires further monitoring and investigation.

It can be concluded that health monitoring in ponds and diagnostic work on fish disease cases in India can provide more than ample material for research on fish diseases and their prevention, thus opening possibilities for incorporation of better health protection measures in aquaculture for higher and more economic yields.

Fish Cell Cultures

These are the essential prerequisite for diagnostic and research work on viral fish diseases: preparing and maintaining of cell cultures requires a special laboratory environment, washing and preparation of glassware, media and other supplies, as well as the application of a technique that is sophisticated for the newcomer to this field. At the end of the consultant's assignment, only improvized conditions for this work could be secured. One attempt to prepare primary cell cultures from rohu could be made and partial success had been achieved. The continuation of this research work and early securing of prerequisites for proper methodology are essential for progress in research and diagnosis of viral diseases.

POND EXPERIMENTS

During the consultant's assignment one pond experiment was completed and the second one started.

In addition to training, the aim of the first experiment was to determine the influence of poor fish handling on losses and the possibility of preventing them by use of antibiotics.

The experiment was carried out in 12 ponds that were randomly used for four experimental treatments of rohu (650 g). Two stocking densities were used. Six ponds (two densities) were stocked with fish that received penicillin plus streptomycin intraperitoneally, while fish for the other six ponds received sham-injections of distilled water. All fish were treated equally otherwise. Their allocation to groups was also randomized. The experiment lasted 12 days. Mortalities occurred 2 to 5 days after stocking and then ceased. All sick and dead fish exhibited symptoms of columnaris disease, while some also had saprolegnia infection of the skin. Flexibacter columnaris was isolated from skin lesions and kidneys of sick and dead fish.

Mortality in fish that received the antibiotics by injection was 1.75 percent and the mortality in non-antibiotic-treated group was 27.18 percent. The stocking density did not affect the mortality rate.

The above experiment indicates that an outbreak of columnaris disease can be triggered in rohu by handling. This is a known fact in fish culture of warmwater species in Europe and North America. The results obtained suggest that injection of suitable antibiotics can decrease losses after handling of broodstock during spawning operations, and that further experiments are needed to assess losses during normal stocking of ponds due to handling, as well as on prevention of such losses by bath or oral treatment of fish prior to stocking.

The second experiment was planned to find out if treatment with a nitraofuran compound can reduce the myxosporidiosis and kidney damage in rohu.

PROPOSALS FOR ORGANIZATION OF WORK

The health of cultivated fish can be adversely affected by numerous biotic and abiotic factors. The number of these factors in fish medicine is even larger than in human or veterinary medicine. Research on fish diseases and fish health protection encompasses, therefore, a constantly increasing number of narrow research fields. As in other research areas, development of sciences and their methodologies now leaves very little space for encyclopaedic oriented type of investigations. Bearing in mind that research on fish diseases has first to gain more ground in India and that practically oriented work must have priority, only a few not quite so narrow fields can be covered at present.

All young scientists starting to work on ichtyopathology and health protection at FARTC should develop knowledge and research techniques on general problems and prevention of diseases. In addition to this, each of them should concentrate on work in one field, according to specific research techniques.

In the first period of research development, general examination techniques for fish, basic bacteriological, parasitological, virological and histopathological techniques, as well as experimentation on prevention and treatments of diseases, should be mastered and used in diagnostic and research work.

Scientists assigned to these fields should work as a homogeneous team, supplementing each other's activity. This is essential for avoiding situations occurring sometimes in some countries with narrow specialists, where the same fish mortality may be attributed to quite different diseases by different researchers, thus confusing the fish farmer or the institution seeking advice for prevention or treatment of diseases.

For the development of working fields it is essential that a significant part of the activity be devoted to diagnostic examination of fish from as many ponds as possible from a large territory. Such activity can be secured through research programmes on health monitoring and taking inventory of disease problems in pond culture. This work is an essential prerequisite for proper selection of research priorities in ichthyopathology, as well as in fields of prevention and treatment. On the basis of diagnostic findings, research can be oriented towards the most urgent needs of aquacultural practice.

Within the diagnostic work, every effort should be made to ensure that fast and reliable diagnostics, together with proper recommendations for disease prevention and treatment reach the fish farmers or owners of the fish as soon as possible. Proper organization of work and development of appropriate methodology are needed for such performance.

Unit for Ichthyopathology and Fish Health Protection

For reasons mentioned in 4.1 above and in order to comply with proposed plans on research approaches and responsibilities at FARTC as the Regional Lead Centre for the project RAS/76/003, the consultant suggests that the four scientists assigned to the work on fish diseases start to function as the Unit for Ichthyopathology and Fish Health Protection. Training and utilization of assigned space has been carried out with this proposal in mind.

Research Fields for Future Planning

Full coverage of fish health problems will be of growing importance for development of aquaculture in India. It is useful therefore to foresee all main research fields which will have to be developed either immediately or at a later stage. These fields may be listed as follows:

i. Fish health monitoring;
ii. Viral diseases;
iii. Bacterial diseases;
iv. Mycoses and algal diseases;
v. Protozoan and methazoan parasitic diseases;
vi. Environmentally caused diseases and stress;
vii. Nutritional diseases;
viii. Intrinsic and hereditary diseases, neoplasms and malformations;
ix. Toxicology;
x. Histopathology;
xi. Prevention and treatment of diseases and their incorporation into technology;
xii. Immunoprophylaxis.

The list may be useful for long-term planning of personnel, space, facilities, training and equipment.

Programmes covering some aspects of fields listed as i, ii, v, x and xi should be initiated immediately, as suggested in subsequent paragraphs of this section and in section 5.

General Diagnostic Work

This basic work should be performed by all four scientists of the unit and should consist of activities described below.

Selecting, packing and transporting fish samples from ponds to laboratory, checking main water quality parameters on the spot and taking water samples. Accepting samples of cultivated carps and taking case histories from extension service personnel, state fisheries department, farmers or staff members of FARTC. Recording arrivals of samples at the laboratory. Keeping records on case histories, findings, diagnosis and recommendations. Preparing annual report on diagnosed cases.

Storage of live or dead fish samples in aquarium, refrigerator or freezer if necessary.

Clinical examination, dissection, macroscopic and microscopic examination of effected and suspect organs and tissues; taking of material for hematological, histopathological, parasitological, bacteriological, mycological, virological and other examinations, as well as for biological experiments (transmission of disease, if necessary for diagnosis). After pathoanatomical diagnosis and collecting data on results of other examinations, setting up of final diagnosis. Quick reporting of final diagnosis and recommendations for prevention and treatment to the owner of examined fish.

Surveys of pond fish populations by sampling for presence of pathogens and diseases throughout the production cycle, with the purpose to identify the health status and eventual needs for preventive treatments. In this way detect specific season - and age - dependent health problems of fish in mono-and polyculture.

Cooperation with scientists in the centre in their work on various aspects of aqua-culture by checking fish health in experiments and recommending prophylactic and therapeutic treatments and advising on health protection measures in planning and conducting research.

Virology

One scientist, e.g., D. Kumar, should be assigned to the work described below.

Registering samples for virological examination, recording of methods applied and of findings, release of virological diagnosis for setting up final diagnosis. Making recommendations for prevention. Preparing of annual report on diagnostic work and research.

Preparation of media and solutions for cell culture work. Subcultivation, maintenance and cryopreservation of existing suitable American and European fish cell lines. Preparation of primary cell cultures from indigenous and other cultivated fish species. Development of fish cell lines from indigenous and other important species for this region.

Extraction of virus and decontamination of inocula for cell cultures or experimental fish. Inoculation of cell cultures, incubation and detection of virus presence. Presumptive and confirmative identification of virus, using morphological, cultural, biochemical and serological characterization. Inoculation of fish, observation of results.

Research on and development of advanced and improved fast virological methods. Use of rabbits for preparing of antisera.

Study and description of viral disease entities by using epizootiological, pathoanatomical, histopathological, virological and serological methods, and biological experiments in aquariums, tanks and ponds.

Surveys of pond fish populations for presence of viral pathogens and diseases in various phases of aquaculture technology. Detection of virus carriers among broodstock and fingerlings.

Bacteriology

One scientist, e.g., K. Suresh, should carry out the work described below.

Registration of samples for bacteriological or mycological examination; recording of method used and findings; quick release of presumptive (if possible also confirmative) microbiological diagnosis for use of setting up final diagnosis, and recommendations for treatment and prevention. Keeping records on recommendations and experiments. Preparing annual report on diagnostic work and research. Preparation of methodology sheets.

Preparation of media and solutions for isolation of various pathogens, determination of their biochemical characteristics essential for classification and diagnosis and for maintenance of isolates. Preparation of stains and staining of smears.

Plating of materials and isolation of pure cultures. Examination of characteristics of isolates (morphology, culture, biochemical and staining characteristics), performing of tests for serological identification.

Research on development of advanced and improved bacteriological diagnostic methods, including immunological and immunofluorescent ones. Study and description of bacterial and fungal disease entities by epizootiological, pathohistomical, histopathological, microbiological and serological methods, and biological experiments in aquariums, tanks and ponds. Determination of pathogenicity and virulence of isolates.

Surveys of pond fish populations for presence of bacterial pathogens and diseases in various phases of aquaculture technology.

Study of normal bacterial flora of species in aquaculture (skin, gills, intestines, blood) and its role in diseases.

Parasitology

One scientist, e.g., B.K. Mishra, should carry out the work described below.

Registration of samples for parasitological examination, recording of findings, quick release of parasitological diagnosis for use in setting up final diagnosis and recommendations for treatments. Keeping records on findings, recommendations and experiments. Preparing annual report on diagnostic work and research. Preparation of methodology sheets (procedures).

Application of various parasitological methods (flotation, digestion, concentration) for detecting protozoan parasites; fixation, staining and taxonomic examination of parasites.

Study of life cycles and ecology of parasites in pond culture, as well as epizootiology of diseases caused by them. Qualitative and quantitative surveys of parasitofauna in various stages of aquacultural technology.

Defining disease entities caused by parasites (epizootiology, histophathology, etc.). Aquarium and pond experiments on these diseases.

Histopathology

One scientist, e.g., R.K. Dey, should carry out the work described below.

Registering samples taken for histopathological examination, recording of methods used and of findings, quick release of histopathological diagnosis for setting up final diagnosis. Preparing annual report on diagnostic results and research.

Preparing of fixatives, histological stains and solutions, shaping up tissues for slicing, slicing by freezing microtome, embedding of samples in paraffin, preparation of paraffin sections, staining, mounting, microscopic examination, case studies and diagnostic work, preparing photomicrographs of sections. Preparation of methodology sheets (procedures). Study of normal histology of various fish species of importance for aquaculture. Study and description of histopathology in individual disease entities caused by various etiological agents (including nutritional and toxic ones). Cooperation with fish nutritionists and other research groups by providing histopathological expertise on status of organs in various experiments.

Treatments

All four scientists should carry out the work described below.

Preparation of lists of drugs for prevention and treatment of fish diseases with indication of source, price and cost of treatment.

Planning and conducting tests on suitability of various drugs for prevention and treatment of diseases in aquariums, tanks and ponds. Keeping records on tests, evaluation of results.

Preparing and giving recommendations on prevention and treatment of diseases in field situations. Keeping records on such treatments and evaluation of results.

Improvement of techniques for treatment of fish by bath, by food and by mass injections.

Collection of Specimens, Pathogens and Slides

Accumulation of documentation for research reports, papers, training purposes and extension work should be a constant activity of scientists in the unit. This, as well as collection, maintenance and preservation of pathogens is also necessary for national and international exchange with fish disease laboratories. A reference laboratory for fish diseases does not exist in this part of the world and the FARTC may assume this role through the unit. Setting up a museum of wet preparations of sick fish, collection of pictures of gross lesions and microscopic in fish, maintenance of fish cell cultures, storage of virus isolates, bacterial pathogens, specimens of parasites, histopathological slides, etc., should be a part of efforts to upgrade the quality of their own work.

Library

The unit should have a small library, consisting of books on methodologies used (virology, bacteriology, parasitology, histopathology, etc.) and on fish diseases. A collection of reprints and a reference card system, arranged according to groups of diseases and related problems, should also be set up at the unit level. Adequate information about a suitable system was suggested to workers by the consultant during the training.

AID IN SETTING UP RESEARCH PROJECTS

Fish Health Monitoring in Composite Fish Culture

Scientists To Work on the Project

One senior scientist (V.R.P. Sinha), specialist for composite fish culture, plus all four scientists of the unit.

Objectives

The objectives are to develop methods of assessing the health status and the occurrence of subclinical and clinical diseases from the stocking to the fishing out of ponds.

Practical Utility

By determining health status and specific disease problems, possibilities will be opened for application of prevention and treatments, thus increasing chances for higher yields. Results of monitoring will also indicate high priority research problems in the field of fish diseases.

Technical Programme

The monitoring should be carried out in three production ponds with high stocking density managed for intensive culture.

Monitoring at three levels will be carried out and compared:

(i) Fish culturist level (farm workers). Daily observation of ponds for fish in distress or dead fish; daily observation of food intake; immediate notification of unit about unusual observations or drop in food uptake.

(ii) Aquaculture specialist level. Extension worker should observe the pond situation at least once every ten days and check fish for growth rate and general appearance by netting at one month intervals. At stocking and at fishing out, the number of fish should be precisely determined to calculate exactly the percentage of loss for correlating this with findings about fish health.

Notification of unit when unusual findings are present.

(iii) Fish disease specialist level. To determine the health status and detect disease or parasite incidence at a 10 percent or higher level, a sample of 30 fingerlings of each species will be examined in the laboratory before stocking and 30 specimens of each species will be examined in the laboratory at bimonthy intervals and at fishing out.

In the unit, the following examinations will be undertaken: total length, weight and condition (Fulton's formula) of each specimen; determination of hematocrit value; dissection, macroscopic and microscopic examination of organs for detecting abnormal appearance of organs and presence of parasites; bacteriological examination of lesions (if they exist) and of kidneys of five fish from each species sample; histopathological sampling of abnormal looking organs, and sampling for virology if symptons suspective for a viral disease exist. Results will be summarized and the situation in the three ponds compared.

Duration of Project

Three years, with extension after necessary adjustments.

PREPARATION AND USE OF PRIMARY CELL CULTURES

Objectives

The objectives are the development of standard methodology for fish cell culture work and virology in cultivated Indian carps.

Practical Utility

It is highly probable that viral diseases will start to occur in fish culture. Their detection, identification and study is impossible without cell culture work. This research is also needed for later development of vaccines for prevention of viral diseases.

Technical Programme

1. Methodology for preparing glassware, media and other items for cell culture work and virology will be modified to suit local conditions;

2. Asceptically taken ovaries and other organs will be used for explant and for trypsinization technique of obtaining monolayers of cells; developing and adopting methods suitable for Indian major carps;
3. Primary cell cultures and foreign cell lines will be kept at various temperatures, subcultivated and maintained;
4. Methods will be developed and standardized for inoculation of cultures with material from diseased and suspect fishes from ponds.

Duration of Project

Three years, with extension after revision.

Development of Methods for Isolation, Identification and Maintenance of Bacterial Pathogens in Indian Major Carps

Objectives

The objectives are the development and improvement of bacteriological research methods for study of fish diseases.

Practical Utility

Diseases have been noted in composite culture that are probably of bacterial etiology. By the above research on suitable methodologies, it will become possible to detect these diseases, to study them, diagnose them quickly and provide the fish culturist with information about prevention and treatment.

Technical Programme

i. Comparison of suitability of various bacteriological media for isolation of various groups of fish pathogens from external lesions (skin, gills) and internal organs (kidney, liver, heart);

ii. Selection and testing of biochemical tests for identification of fish pathogens;

iii. Development of immunological methods for identification of bacterial pathogens, using rabbit hyperimmune sera;

iv. comparison of suitability of several commercially available or other media for maintenance of isolates.

Development of Methodologies for Detection and Identification of Diseases Caused by Parasites

Objectives

Sensitive and reliable methods for detecting and identifying internal protozoan and methazoan parasites in Indian major carps will be developed.

Practical Utility

Information on diseases caused by parasites is restricted in India mostly to skin and gill pathology. It is, therefore, necessary to develop methods for early detection, concentration, storage and staining of potentially harmful internal parasites. These methods will enable carrying out properly the fish health monitoring and expanding research on diseases damaging fish culture, thus opening possibilities for prevention and treatment.

Technical Programme

i. Blood parasites - development of techniques for taking blood, preparing smears and staining them for blood parasites;

ii. Intestinal cocedia - detection, concentration and staining of spores;

iii. Myxosporidia - detection, digestion methods, concentration and comparison of various staining methods for myxosporidia in kidneys and other organs.

Histopathological Methods for Study of Kidney and Liver Pathology in Indian Major Carps

Objectives

The objectives are to determine the suitability of major histopathological staining methods for study of pathological changes in two major organs of carps.

Practical Utility

Most systemic bacterial and viral diseases of fishes, some localized parasitic diseases and nutritional deficiencies manifest themselves by histological alterations in these two organs. The comparison of results obtained by various staining methods on material from large number of fishes will allow choosing standard stains for examining these organs for routine and specific purposes.

Technical Programme

i. Organs collected from 40 healthy and 40 sick specimens of each species of Indian major carp will be fixed by buffered formalin and Bouin's fluid;

ii. Paraffin and freeze - sections of organs will be stained by hematoxilin-eosin, PAS, PAS alcain-blue, Malloxy, Sudan black and Masson's trichrome stain;

iii. Stained sections will be evaluated for suitability to study normal and altered histology of kidney and liver.

2

India Fish Genetic: Selection and Hybridisation

INTRODUCTION

The Government of India, assisted by the United Nations Development Programme and the Food and Agriculture Organization of the United Nations, are engaged in the Intensification of Freshwater Fish Culture and Training Project (FI:DP/IND/75/031) at the Freshwater Aquaculture Research and Training Centre (FARTC) of the Central Inland Fisheries Research Institute (CIFRI) at Dhauli, Bhubaneswar, Orissa. As part of the operation of the project, FAO assigned Dr J. Bakos as a Fish Culture Genetics Consultant for three months from 29 October 1980 with the following terms of reference:

1. To assist in establishing the fish genetical laboratory and in elaboration of studies on the selection of fish strains with better growth potential, meat quality, resistance to disease and longer breeding period;
2. To assist in the development of effective gynogenesis technique for carps;
3. To assist in identification and maintenance of land races of carps towards development of genetic pool.

During the assignment the consultant had detailed discussions with Dr V.R.P. Sinha, Project Director and Head of FARTC; Mr R.D. Chakrabarty, Head of the CIFRI Substation, Cuttack, and other scientists involved in breeding and genetical work. The Consultant wishes to thank all of them for their fruitful discussion. Thanks are also due to

Dr Maharana of Cuttack Medical College for providing the cobalt 60 gamma ray instrument for the work.

Fish Species Cultivated in India

The following fish species are commonly cultured in India and are thus the most important from the point of view of genetic selection and hybridization.

Indian major carps:

- Catla (Catla catla H.)
- Rohu (Labeo rohita H.)
- Mrigal (Cirrhinus mrigala H.)

Exotic carps:

- Silver carp (Hypophthalmichthys molitrix Val.)
- Grass carp (Ctenopharyngodon idella Val.)
- Common carp (Cyprinus carpio L.)

Other fish species which are important for fish culture are as follows:

Indian minor carps:

- Reba (Cirrhinus reba H.)
- Bata (Labeo bata H.)
- Calbasu (Labeo calbasu H.)
- White carp (Cirrhinus cirohosa Bloch)
- Cauvery carp (Labeo kontius Fedon)

The Indian major carps are the endemic species in India and their contribution to the total fish production is considerable. Thus, it is essential to improve their genetical capabilities and also to maintain their gene bank. Equally important are the Chinese carps, which also need proper genetical selection.

SUMMARY DESCRIPTION OF WORK UNDERTAKEN

Establishment of the Genetical Laboratory

The Genetical Laboratory consists of two rooms, one of them air-conditioned for under-taking biochemical investigations, such as electrofocussing, etc. The other room is for chromosomal investigations, gynogenesis and other genetic experiments. Another room will serve as a laboratory for selection and hybridization work. A fourth room will have provision for aquaria for rearing and keeping the progeny after gynogenesis and sex reversion. Here necessary equipment will be kept for marking, measuring and dissecting the fish. This room needs some drainage facilities for about 15 fibreglass tanks of different sizes.

After carefully examining the requirements of the genetical laboratory and the equipment already available, listed below is the additional equipment needed for the laboratory:

Item	Quantity
(1) Gel electrophoresis apparatus (both disc and slab)	1
(2) Densitometer	1
(3) Spectrophotometer	1
(4) Microscope, Dialux 20	2
(5) Centrifuge, 500–5 000 rev/min	1
(6) Centrifuge, refrigerated 10 000–13 000 g	1
(7) Camera	1
(8) Dissolved oxygen meter, electronic	1
(9) Microscope, Olympus	1
(10) Microscope, Stereo WILOM$_3$	1
(11) Refrigerator 265 1	1
(12) Ultra-violet germicidal tube, 30 W Philips with extra tube	1+1
(13) Refrigerator-deepfreezer 100 1	1
(14) Refrigerator-deepfreezer 500–1 000 (-20°C)	1
(15) Meat mincer	2
(16) Electronic mixer	2
(17) Fibreglass tanks 1 000 1, 500 1, 200 1 capacity	10 each
(18) Electric aerator	5
(19) Balance 0.5 g – 1 000 g	1
(20) Dia slide projector	1
(21) pH meter, electrical Philips	1

Pond Requirements

Five different land races of catla, rohu, mrigal, silver carp, grass carp and common carp have to be maintained separately using a polycultural stocking system. As a safety measure, some specimens from all fish species must be maintained in two ponds in mixed conditions and permanently marked. This work requires 32 ponds (60 m^2).

The fish fry produced with artificial propagation should be reared separately. The expected number of different populations yearly (three species with five various land races need 15 ponds of 150 m^2), one intraspecific hybridization with polyallel crossings of five different females and males would need 25 ponds of 150 m^2 and, for intergeneric hybridization, 5 ponds of 150 m^2. Thus, altogether 45 ponds, 150 m^2 each would be required for fry/fingerling rearing.

The area of testing ponds should be 2 000–5 000 m^2 each. About 30 such ponds will be required for undertaking experiments with land races, hybrid populations, etc.

Further, about 10 ponds, 60 m^2 each may be required for other genetical research. Therefore the total requirements of the ponds would be:

1. 42 ponds of 60 m^2
2. 45 ponds of 150 m^2
3. 30 ponds of 2 000–5 000 m^2

Development of Gynogenesis Technique for Carps

Because the major carps and Chinese carps breed only during the monsoon season, common carp was selected to be the test fish (common carp also breed during December-January) and carefully selected males and females were induced to breed with pituitary hormone injections. The sperm was irradiated with cobalt 60 and also some by ultra-violet ray. Stickness of the eggs was removed by urea and salt solution. During the embryonic development of eggs, a proper check was made of the development of the gynogenetic progeny. Identification was made of haploid and diploid embryos.

In certain countries a cobalt 60 gamma ray source in 100 000 Rad dose is given for irradiating the sperm. However, because of the difficulties in cooling the sperm here and also the duration of the cobalt 60 gamma ray, experiments were conducted to irradiate the sperm with the help of an available ultra-violet lamp. Appropriate duration of irradiation, intensity of the exposure and dilution of sperm were not known, so experiments were carried out to determine these. It was found out that the irradiation time should be 60 min, distance between the lamp and the sperm to be 12 cm, dilution 1:2 (dilution with frog Ringer) and the thickness of layer 2 mm. Mechanical shaking was used during the time of irradiation with a frequency of 120/min. Further experiments are necessary to find out the suitable method for cooling of sperm to 3–4°C during the irradiation.

The rearing of gynogenetic common carp progenies was carried out in plastic pools. They were initially given zooplankton collected from fish ponds for 28 days. After that they were given balanced artificial food consisting of sheep liver and heart and broiler chicken feed supplemented with vitamin A, B, C, D and E.

The fry showed better growth rate than the normal fry of common carp. In order to have sex reversal in these fry, they have been fed with feed containing methyltestosteron. The result of the investigation will be known in due time. The above technique of gynogenesis has been carried out in collaboration with the counterpart scientists assigned to the consultant. Thus, these scientists are fully conversant with and trained to utilize this technique in other carps during their breeding season.

Identification and Maintenance of Land Races of Carps

Collection of various land races of cultivated carp species (preferably at the fingerling stage) from different water systems of India is in progress, the first consignments of fish populations was to arrive in January 1981. Those carp species are also to be collected from Cuttack. Together with the Cuttack strain, five different land races of each species are suggested to be collected. FAO, in collaboration with the United Nations Environment Programme (UNEP), expects to develop a programme to preserve the genetic resources of aquatic organisms. Therefore, it would be better to have a link of the collection and maintenance of land races of carp along with the above programme.

After collection and arrival of fingerlings they need to be stocked in separate ponds. The various land races, as basic populations for research, have to be marked with fin clipping and operculum tagging. They should be reared in optimum conditions so as to achieve their best growth rate and early maturity. It is necessary to investigate the morphological/biochemical characteristics of different populations.

CURRENT RESEARCH PROJECTS FOR FISH GENETICS

Selective Breeding and Hybridization

Studies on cytogenetical features of carp hybrids improving the strains of Indian major carps with crossing 1977–82.

Breeding of selected stock of grass and silver carps. Improved the stocks of Chinese carps 1977–81. Studies on the morphology, growth and maturity of the hybrid between grass carp female and silver carp male 1978. Hybridization between Labeo rohita × Cyprinus carpio; C. mrigala × C. Carpio; C. catla × C. carpio and C. catla × H. molitrix 1977–80.

In the field of genetics CIPRI attempted some preliminary experiments in interspecific and intergeneric hybridization, mainly for the purpose of improving the market quality of the fish (smaller head and deeper body, as in crosses between catla and rohu) and obtaining faster growth rates.

Even though the results achieved are encouraging, it has not yet been possible to carry out adequate progeny testing due to a lack of facilities at the Cuttack farm. Genetic selection experiments to prevent inbreeding depression or to obtain strains with greater resistance to temperature variations common in tropical ponds, resistance to diseases and enhanced growth rates have not been attempted.

RECOMMENDED RESEARCH PROGRAMME AND ITS POSSIBILITIES

Methods and Phases of Research Work

Comparative Testing of Fish Species and Different Lines

The aim of comparative testing is to determine and compare the most important characteristics which are closely correlated with the productivity of species. The different phases of comparative testing are as follows:

1. The groups for testing have to be produced with artificial propagation. One female and one male can be used which are chosen by phenotypical selection with the best growth rate and appearance.
2. The groups of progenies will be reared in separate ponds and after six months they need to be stocked together in a common pond with group marking (150–200 fish from each population at least). The stocking rate of testing ponds must be the same as in commercial fish farming/intensive polyculture. When the number of stocked fish from one population are less than 100 fish, the testing work has to be repeated three to four times with the same groups of fish.
3. It is necessary to use organic and inorganic fertilizers to fertilize the water and also to feed the fish.
4. The fish population reaching a marketable size need to be harvested. Proper measurement of every individual is required for evaluating different characteristics.
5. The qualitative characteristics, such as the colour, scales and the formation of fins are not closely correlated with the productivity, therefore the evaluation of those only for the purpose of morphological characterization of the populations has any importance.
6. From the quantitative characteristics one has to evaluate the viability, growth rate, food conversion, quantity of meat, quality of meat (protein and fat content) and the reaction to fish diseases. Later, by producing inbred lines, the fecundity and the oxygen demand tolerance also has to be examined.
7. The results obtained need to be verified with variance analysis.
8. The productivity of various groups can be compared through a multifactorial, "100 score" system by the four most important characteristics (like viability, growth rate, food conversion, quality of meat).

9. If the test is arranged in more ponds with replicates, standard population control has to be used in all ponds, which gives the possibility to compare the differences of environment among the ponds used. The common carp is suggested as a standard control considering its omnivorous feeding habits. The standard control can be produced in every year with a panmictic propagation, maintaining their quality and characteristics on the same level during a long time.
10. The results of pond testing experiments can be compared with the short laboratory tests and with the results of oxygen demand toleration (ODT) tests. The average of the above three methods gives a reliable final result of comparative testings.
11. The outstanding populations showing best characteristics will be used for the further selection and hybridization work.

Progeny Testing for Determining the Individual Breeding Value

The progeny testing is a suitable method to determine the individual breeding value of males and females after the productivity of their progenies. The phases of the work are the following:

1. For the evaluation of males, the eggs of two females have to be fertilized with the sperm of chosen males separately, producing maternal half sister groups. Five or six males can be examined at the same time in their ten or twelve progeny groups from the two different females. For the evaluation of females one must follow the same system, using two different males.
2. The productivity of half sister groups are examined and compared by using the above mentioned comparative testing method (i.e., 4.1.1).
3. The average productivity of two groups originating from different females but from the same male shows the individual breeding value of the given male fish.
4. The best males and females in breeding value can be used for the intraspecific hybridization, or for improving the species in their characteristics.

Intraspecific Hybridization

The aim of intraspecific hybridization is to produce more productive hybrids by crossing different genotypes of females and males having the best results during the previous progeny testing. The phases of intraspecific hybridization are as follows:

1. Selection of various genotypes of the same species;
2. Inbreeding of the selected lines;
3. Polyallel crossing of the different lines;
4. Comparative testing of the F_1 hybrid generation;
5. Propagation of the best hybrid producing parental lines for the fish seed production farms;
6. Producing three-line hybrids;
7. Producing four-line hybrids.

After the inbreeding of parental lines, the heterosis effect can be increased. The fastest method of inbreeding is the artificial gynogenesis.

As a result of intraspecific hybridization, the increasing productivity of fish species of about 15–25 percent can be expected.

Artificial Gynogenesis and Hormonal Sex Reversion

The main purpose of gynogenesis is to produce inbred individuals having only maternal chromosomes for the further genetic work. For causing gynogenesis artificially, inactivated sperm can be used for fertilization of eggs, with a cold shock treatment after fertilization. For the inactivation of sperm a cobalt 60 gamma ray source or ultra-violet germicide lamp has to be used. The individuals of gynogenetic populations are females with diploid chromosome sets similar to the mother. The technique of artificial gynogenesis was demonstrated on the common carp. The method can be adopted for the other fish species with necessary modifications.

For producing gynogenetic inbred male populations, methyl-testosteron hormone feeding has to be provided for the females at a young age to influence the sex formation in the direction of the male. The technique of sex reversion was demonstrated with gynogenetic common carp. The same method can be used for all fish species which are able to feed with artificial food. The inbreeding coefficient value (F) of the first gynogenetic population is already 0.90, which value can be reached only after 12 generations of sister-brother pairing.

For the purpose of increasing the inbreeding level, the gynogenesis can be applied from generation to generation. It is important to make a proper selection within the gynogenetic population to avoid the undesirable effect of inbreeding depression. The keeping and maintenance of gynogenetic inbred females and males requires more care than the normal ones because of their high breeding value and more sensitiveness to the environmental effects. Therefore, the location, rearing and keeping of this fish in small ponds is necessary, with a good water supply and other necessary conditions and to protect them from poaching. Some outstanding individuals can be kept in an indoor raceway system. The gynogenetic inbred populations or individuals can be used for improving the species with their useful characteristics, to make inbred lines for intraspecific hybridization, for producing monosex fish populations and also for genetic examinations for particular purposes.

Application of Gynogenesis for Improving Productivity of Fish Species

Producing and maintaining inbred lines for intraspecific hybridization: The greatest advantage of producing inbred lines is to obtain higher heterosis effect after the crossing of gynogenetic female and male lines.

The main phases of the work are as follows:

1. To select the best females for performance and morphological features.
2. Producing the first gynogenetic female generation (G_1)
3. Rearing the gynogenetic fry suitably for producing sexually mature fish as soon as possible.
4. Selection by phenotype for the best growing and correct body formation.
5. Producing the second gynogenetic female generation (G_2).
6. After dividing the G_2 population into two parts, one part will be reared as a female group and the other part has to be oversexed for producing gynogenetic males (G_2).

7. Rearing the G_2@& and G_2 groups suitably.
8. Brother-sister pairing can be used for producing the third gynogenetic generation (G_2@& × G_2 — G_3@&). The purpose of brother-sister pairing is to avoid the chance injuring of chromosomes.
9. Sex reversion in the G_3@& progeny group.
10. Brother-sister pairing for producing fourth gynogenetic generation (G_2@& × G_3 — G_4)

This scheme can be continued by the previous pattern. During all the phases the best female or male are selected according to their better constitution.

The inbred population produced can be used after phase five for intraspecific hybridization. After top crossing, good heterosis effect can be expected.

Producing Semi-gynogenetic Lines

The purpose of this work is to produce gynogenetic inbred male lines in cases when the hormone feeding is different, such as the case of silver carp and catla. The phases of work are as follows:

1. From population "A" selection to be made of the best females.
2. Producing the first gynogenetic generation (AG_1@&).
3. Crossing the AG_1@& female with a normal AB& male. The result obtained will be a semi-gynogenetic bisexual population (50% @& and 50% B&).
4. Selection for the best females and males.
5. Producing the second gynogenetic generation (AG_2@&)
6. Crossing of the AG_2@& female with the normal AB& male. The scheme can be continued by the previous pattern. After the third phase the gene pool of population becomes homozygotic to about 60 percent.

Producing Monosex Fish Populations with Gynogenesis

The purpose of this work is: (i) to produce monosex female populations for pond fish culture because of their faster growth than that of the bisexual group of the same age; (ii) the stocking of monosex populations in natural waters like reservoirs or undrainable lakes; (iii) to develop inbred gynogenetic male lines is necessary for maintenance only in commercial fish producing farms or for fish seed production farms and pairing them with normal females; (iv) after solving the problem of deep freezing the sperm, one can collect and store the sperm of inbred gynogenetic suitable males, so as to use it for producing monosex populations in other farms. The phases of the work are as follows:

1. To select female from a male line
2. Producing the first gynogenetic population (G_1 @&)
3. Producing a gynogenetic male population with hormonal sex reversion (G_1)
4. Using the selected males for top-crossing by pairing them with normal females.

The progenies will be all females which may show improved characteristics.

Intergeneric Hybridization

Intergeneric hybridization consists of crossing fish belonging to different species. The aim of producing intergeneric hybrids is as follows:

1. To unite the useful characteristics of two different species in one of the new forms.
2. Producing a new construction in consequence of the heterosis effect..

Using intergeneric hybridization, the following results can be expected:

1. New construction with new feeding habits for composite fish culture or for water reservoirs
2. Better growth rates and survival rates
3. More restful behaviour, having better tolerance for reduced levels of oxygen (silver carp × bighead)
4. Better quality of meat (common carp × silver carp; rohu × catla)
5. Monosex triploid grass feeder fish (grass carp × bighead)

6. Better tolerance of water salinity for brackishwater fish culture.

In the field of intergeneric hybridization the scientists of India have achieved good results (see Section 3).

The phases of intergeneric hybridization are the following:

1. Studying the fish species characteristics; behaviour and genetic structures
2. Analysis of chromosome sets and karyotypes
3. Crossing the species, using artificial propagation
4. Rearing the hybrids in the most suitable environment for their expected habitat
5. Comparative testing of hybrids and the parental forms
6. Selection of the most suitable individuals
7. Artificial propagation experiments on the F_1 hybrids
8. Stabilizing the new characteristics with inbreeding or back-crossing with the closer parental species which are similar in their main characteristics to the new hybrids
9. Describing the new hybrids' morphological, biological and economic characteristics.

Other Research

Biochemical Polymorphysm Investigations

These investigations are aimed at finding and using genetic markers for characterization and identification of fish populations and individuals. Such markers are the enzymes and proteins, including transferrins. The method is very important for characterization of species, land races and hybrids. Also this is useful in testing the success of gynogenesis in fully scaly fish species. The instrument for this method is the polyacrilamid gel-electrophoretic equipment which was ordered.

Chromosome Investigations

These investigations have the aim of characterization of fish species and hybrids to study the possibilities of intergeneric hybridization. The method of chromosome investigation is well known by the scientists but some technical details have to be improved. These investigations will be also important in the future research programme with the aim of polyploidization.

List of Research Subjects Recommended for the Next Ten Years

1. To develop practical methods of artificial gynogenesis and hormonal sex reversion of catla, rohu and mrigal to produce highly inbred parental lines. Intraspecific hybridization of catla, rohu and mrigal, with crossing gynogenetically inbred female and male lines, to compare the productivity of the different populations.
2. To develop practical methods of artificial gynogenesis and hormonal sex reversal of silver carp and grass carp to produce highly inbred parental lines. Intraspecific hybridization of silver carp and grass carp with crossing gynogenetically inbred female and male lines. To compare the productivity of the different populations.
3. To develop practical methods of artificial gynogenesis and hormonal sex reversion of common carp to produce highly inbred parental lines. Intraspecific hybridization of common carp with crossing gynogenetically inbred female and male lines.
4. Fish chromosome investigations on Indian major carps and cultivated exotic carp species.
5. Biochemical polymorphism investigations on Indian major carps and the cultured exotic carp species to determine the transferrin and isoenzyme polymorphysm of different genotypes with a view to selection and hybridization of cultivated species. Intergenetic hybridization among:
   a. Indian major carps and minor carps
   b. Exotic carps
   c. Indian major carps and exotic carps.
6. To develop practical methods of producing polyploid and androgenetic fish populations in Indian major carps and the common carps.
7. Establishment and maintenance of the living gene pool of

Indian major carps, collecting the different land races, elaborating practical marking systems, propagating for renewal of populations every five years and using them in the genetic and selection work.

8. Spreading the new fish hybrids into the commercial fish farms and fish seed producing farms in the form of inbred female and male lines.

EXPECTED RESULTS OF THE PROJECT RESEARCH

Some results of project research will provide a scientific basis for future research but other results should assist the producing sector to develop better results in actual fish farming. As a result of chromosome and biochemical investigations one can become acquainted with the genetic structure of cultivated fish species and obtain additional information to determine the further directions of the work of selection and hybridization.

By the comparative testings and progeny testings the best populations and individuals are selected in terms of their productivity. Using the best strains, the population can be improved, not only in the Institute but also in the fish farms. By improving productivity of species with intraspecific hybridization, on the basis of Hungarian experience, the following results can be expected: in the survival rate + 20%, growth rate + 20%, conversion of food + 15% and increasing percentage of eatable meat + 2–3%.

The monosex female populations provide 10–15% higher growth than the bisexual mixed ones in the same environment. By artificial gynogenesis and hormonal sex reversion inbred female and male lines can be produced, and from their heterosis hybrids 30–40% higher productivity can be expected. By improving the fish species by the above mentioned methods, the productivity of composite fish culture should be increased about 15–20% without any extra financial investment.

RECOMMENDATIONS

Rearing of the highly valuable fish fry produced with gynogenesis and sex reversion can be accomplished only in overflow, raceway systems. The presently used plastic pool system is unsuitable for keeping and feeding fish for long periods. Therefore, it is necessary to transform the fish aquarium laboratory into a modern overflow aquarium room urgently.

One of the basic works of genetics and selection is the artificial propagation of selected fish. FARTC has a plan to build a modern, multipurpose fish hatchery. The hatchery should be constructed as soon as possible to enable many important research projects on genetics and breeding to be undertaken.

It is recommended to import and acclimatize the Hungarian strain of common carp, which has been selected for its good growth rate, good food conversion and longer period of sexual maturation. It would be important to examine this common carp in composite fish culture conditions.

3

A Strategic Assessment

Background

Against a background of stabilizing or even falling catches from traditional capture fisheries, increasing human population and an ever-increasing global demand for food, there is now a major challenge in re-examining the nature and potential of the aquaculture sector, and to ask just how far aquaculture might go to satisfy the expected increase in demand for its products.

Aquaculture is a notably diverse activity, and an appreciation of this diversity is fundamental to the understanding of the issues influencing its future development. This applies to an assessment of requirements for resources of all kinds, to the way aquaculture may develop independently from the fisheries sector, to the prospects for producers from all economic levels or locations, or to opportunities for developing new markets (Muir, 1995).

The present study is an attempt to satisfy part of the need for information on aquaculture potential in Latin America. This study builds on the experience gained in Africa in the analysis of factors important for aquaculture development and operation. (Kapetsky, 1994) and the implications of aquaculture development for food security (Kapetsky, 1995).

At the same time, it integrates a growth model to improve the estimates of fish yields in order to estimate the potential for warm-water and temperate-water fish farming in the inland waters of Latin America.

Objectives

This study was conducted for two purposes:

- to stimulate improved planning for aquaculture development at the national level; and
- to plan comprehensively for technical assistance activities by FAO and by other national and international organizations that provide both technical and financial assistance for aquaculture development.

STRATEGIC ASSESSMENTS OF AQUACULTURE POTENTIAL

This study was carried out using a geographical information system (GIS) for the basic analyses. A GIS is an integrated system of hardware, software and personnel that is used to manipulate, store, analyse and report spatial (geographic) data. By allowing manipulation of large spatial data sets and providing a rigid analytical framework, integration is possible of diverse knowledge that may be both qualitative and quantitative in nature, such as soil type, estimates of water requirements and yield potential by the use of models, etc. By scrutinizing Latin America for fish farming potential in relatively small units (i.e., grid cells of approximately 9 km × 9 km), GIS technology provides a means to ask "Where is the potential, of what quality and over how much surface area?"

GIS is being increasingly employed in aquaculture. An overview of GIS methodology appropriate for aquaculture is provided by Meaden and Kapetsky (1991) and a review of applications was made by Kapetsky and Travaglia (1994). Most applications in aquaculture are for site selection over rather limited geographic areas, but strategic assessments of aquaculture potential have been made for larger areas, such as countries by Kapetsky *et al.* (1991) and states (Kapetsky, Hill and Worthy (1990); Aguilar-Manjarrez and Ross, (1993); Aguilar-Manjarrez (1996)). However, the only continental-level strategic assessment of aquaculture potential is that by Kapetsky (1994).

The GIS approach can be complementary to other approaches, such as that used by Muir (1995) to estimate aquaculture potential worldwide on a country by country basis using income, resource and development indicators, and by Born, Verdegem and Huisman (1994), who looked at macro-economic factors that influenced world aquaculture production from one region to another.

Strategic assessments of regional pond aquaculture potential require estimates of fish yields that are possible at different geographical locations. In the above-mentioned study in Africa (Kapetsky, 1994), fish yields (expressed in terms of crops/y) were estimated on the basis of temperature thresholds established for the model species (Nile tilapia, *Oreochromis niloticus*).

However, this approach is not readily extendible to other species that may be of interest. Moreover, the approach does not directly consider the effects of seasonal water temperature variation on fish growth and food consumption rates, nor does it account for the effects of other factors (e.g., feeding levels, photoperiod and fish size) on these rates. Bolte, Nath and Ernst (1995) have developed a fish growth model that accounts for the effects of all of these factors on fish weight, and therefore yields. Techniques have also been developed to automatically adjust the parameters of this model for different fish species (Bolte and Nath, 1996).

Water temperature is an important input variable required for regional-scale assessment of fish yields. One approach to water temperature involves the use of heat balance models (e.g., Fritz, Meredith and Middleton, 1980). Such models account for the effects of geographical variations in air temperature, and in other weather characteristics (solar radiation, cloud cover, wind speed and relative humidity) on pond water temperature. Nath (1996) has developed a heat balance model for use in pond aquaculture.

Both the heat balance and growth models cited above have been packaged in the decision support system POND ©[1] (Bolte, Nath and Ernst, 1995; Nath, Bolte and Ernst, 1995), which runs under the Microsoft Windows operating system.

1 *POND* is a copyright of Oregon State University. It has been developed as part of the activities of the Pond Dynamics/Aquaculture Collaborative Research Support Program (PD/A CRSP), which is supported in part by the U.S. Agency for International Development.

INLAND AQUACULTURE PRODUCTION IN LATIN AMERICA

Inland aquaculture production in Latin America is rising, but in absolute terms it is small in relation to production from inland fisheries and somewhat less than mariculture production. For the most recent year for which information is available (1994), inland aquaculture amounted to 120 000 tonnes, while mariculture provided more than twice as much, 286 000 tonnes. Inland capture, at 500 000 tonnes, exceeded both kinds of aquaculture. In comparison, marine fisheries production was about 23 million tonnes.

SMALL-SCALE AND COMMERCIAL AQUACULTURE IN INLAND LATIN AMERICA

Two types of rural aquaculture in Latin America are discussed by Martinez-Espinosa (1994, 1995). Type one is the "poorest of the poor" or subsistence aquaculture, which is characterized by very low cost accompanied by very low output. Producers who have excess production sell or barter it locally. The second type is "less poor" with low to medium costs and low or medium output. This type of aquaculture is conducted by farmers who add aquaculture to traditional agricultural activities. Most of the production is sold. A third type, "industrial", can be a corporate, high-investment, stand-alone activity.

Of the various types, it is the low to medium investment, low to medium output type that apparently has excellent potential.

This study estimates potential in three categories that relate well to the above-described situation: small-scale fish farming and two levels of commercial farming. The first of these, small-scale, corresponds to very low investment, very low output. The first level of commercial farming is semi-intensive and it includes the farmer who would add aquaculture to traditional farm activities. The second level of commercial farming is intensive, and pertains to those who would undertake aquaculture at an industrial level.

METHODS

Methodological Overview

This investigation is similar in many ways to a traditional site selection study for inland fish farming. That is, the study examines how well sites satisfy the physical and economic requirements of the culture system and how well four candidate species will perform at the sites under small-scale farming and two levels of commercial farming. It differs only in that a GIS is used to quantify the potential and in that the study is synoptic for all of continental Latin America. Each "site" that is being evaluated is an approximately 9 km × 9 km (at the equator) grid cell, of which there are nearly one-quarter million in Latin America.

Two important constraints were placed on this study in order to save on time and costs: only already digitized (i.e., computer-ready) maps were employed for the analyses, and field verification studies were limited to one country. An underlying objective was that the results had to be comparable for every area and every country of the continent. This imposed an additional constraint on the data: they had to be comprehensive for all of the continental countries. Most unfortunately, due to the lack of one gridded data set - mean monthly precipitation - it was not possible to include the Caribbean island countries in this study.

Throughout this study, four levels of suitability have been used in order to keep the analyses manageable and to make the results more easily comprehensible and comparable. The levels are very suitable (VS), suitable (S), moderately suitable (MS) and unsuitable (US). The interpretation is, in general, that the VS level provides a situation in which minimum time or investment is required in order to develop fish farming. For a level classified as S, modest time and investment are required, while if MS, significant interventions may be required before fish farming operations can be conducted. If the suitability level is US, the time or cost, or both, are too great to be worthwhile for fish farming.

The study was carried out in five steps:

- Each grid cell was analysed as to its suitability for the culture system - farming fish in levee ponds - based on a combination of five physical and economic factors.
- The importance of each factor was established relative to all of the others as commercial and small-scale fish farming models.
- Each grid cell was analysed as to its fish production potential

for four culture species at two levels of commercial farming and for small-scale farming using a yield model based mainly on water temperature, feeding levels and weights at harvest.

- The commercial and small-scale site suitability models were used together with the species yield model to reach a combined evaluation of the suitability and yield of each grid for two levels of commercial fish farming and for small-scale fish farming.
- The results were verified.

In more detail, for the first time known to us, a bioenergetics model was incorporated into a GIS to predict the number of fish crops possible per year over a large geographic area. A gridded temperature data set was used as input to the bioenergetics model to predict numbers of crops per year for four species: Nile tilapia (*Oreochromis niloticus*), tambaqui (*Colossoma macropomum*), pacu (*Piaractus mesopotamicus)* and carp *(Cyprinus carpio)*. Among these species, Nile tilapia is being increasingly cultured in the warmer waters of Latin America because of its ease of culture, rapid growth rates, and tolerance to a wide range of water quality conditions. Carp is currently not widely cultured in Latin America. However, this species has a wide range of temperature tolerance and the potential for its culture is thought to be high. The characids (tambaqui and pacu) are commercially cultured in several countries (e.g., Brazil, Colombia and Venezuela) and are generally good candidates for pond aquaculture (Saint-Paul, 1989; Lovshin, 1995). Tambaqui generally performs well in warm waters (exceeding about 20°C) whereas pacu is more tolerant of colder temperatures.

By varying input levels and size at harvest, the performance of each species under two levels of commercial fish farming and for small-scale fish farming was evaluated for each grid cell. In addition to the suitability of each 9 × 9 km grid cell for the production of the above-mentioned species, each grid cell was evaluated for a number of other factors important for fish farm development and operation. These included:

- urban market potential based on travel time proximity and population size of urban centres,
- potential for farm-gate sales, based on population density,

- engineering and terrain suitability for pond construction, using a variety of soil attributes,
- water loss from ponds due to evaporation and seepage, and
- availability of agricultural by-products as feed and fertilizer inputs, based on crop potential.

Commercial and small-scale aquaculture models were developed by weighting these factors using a multiple criteria evaluation procedure. Areas unavailable for inland fish farming development were identified by incorporating protected areas, large inland water bodies and urban centres as constraints. Finally, the suitability of each grid cell for each of the four species was analysed together with the other factors through the commercial and small-scale models to identify those areas of the continental Latin American countries that ranged from very suitable to moderately suitable for inland fish farming at commercial and small-scale levels.

GIS PROCEDURES

This study is aimed at individuals concerned with aquaculture. Therefore, this section contains only information sufficient to convey the analytical procedures and the rationale of the methods employed.

Factors for the Development and Operation of Fish Farms

In all, five factors were considered in order to assess potential for commercial farming: urban market size and proximity; farm-gate sales; soil and terrain suitability for ponds; water loss; and agricultural by-products as inputs. All except urban market size and proximity were taken into account for small-scale farming. The above factors are continuous in nature. Another set of constraints included in this study, such as reserved areas, exclude some geographical areas altogether. These constraints are dealt with at the end of the section.

Urban Market Size and Proximity

Urban market potential pertains only to commercial fish farming in this study. It is assumed that small-scale fish farmers consume part of their production and either sell or barter the surplus locally. Urban market potential was defined using ideas of travel time proximity and market size based on population.

Time Proximity

In terms of time, it was assumed that a 12-hour round trip from the farm to the market centre was the maximum possible in order to preclude the cost of the driver (and helper, if any) staying overnight away from the farm.

Time proximity was defined using the road system. Because of an oversight in classifying roads in the original data obtained from the Digital Chart of the World, Version 1 (ESRI, 1992a), it was necessary to assume that all roads were motorable and all-weather. A speed of 90 km/h was assumed for a truck travelling over these roads. Further, it was assumed that within each grid cell with a road system there also would be a feeder road system, not mapped, and that the truck travelled at one-fourth as fast as over the mapped road system (22.5 km/h).

Using the algorithm described in Annex 1, it was then possible to spatially define main and feeder road travel time combinations in four classes, with up to a maximum of 6 hours, or less, in one-way proximity.

Market Size

Urban areas were defined using ARC/WORLD 1:3M data (ESRI, 1992b) and its accompanying classification system. The population data were re-classified in order to make four classes.

Table: Classes of market potential based on population

Suitability	Market Potential Classes	Original ARC/WORLD 1: 3M Classes
	(inhabitants/km^2)	(inhabitants/km^2)
Very Suitable	> 1 000 000	> 5 000 000
		1 000 000 – 5 000 000
Suitable	250 000 – 1 000 000	500 000 – 1 000 000
		250 000 – 500 000
Moderately Suitable	50 000 – 250 000	100 000 – 250 000
		50 000 – 100 000
Unsuitable	< 50 000	< 50 000

Time Proximity and Market Size Combined

The two sub-factors, travel time and market size, were combined to make a consolidated factor "urban market size and proximity," whereby each of the four urban size classes were combined with the four classes of time proximity and interpreted as urban market potential.

FARM-GATE SALES

Market potential as farm-gate sales was inferred from population density (in individuals/km^2 or i/km^2) because no comprehensive marketing data were available on a scale compatible with that of this study.

Population density data were obtained from the National Center for Geographic Information and Analysis (Tobler *et al.*, 1995). The basic assumption was that the greater the population density, the better the market opportunities for farmed fishes (Kapetsky, 1994). A second assumption was that there is a relationship among population density, land use and land cost so that at some point alternative uses of the land preclude its use for aquaculture. Thus, relatively high population densities become a constraint to the placement of a fish farm because of the increased probability of high land costs. For small-scale aquaculture that is pursued as one of a number of activities on a single-family farm, the density at which alternative uses displace farming is probably related to sub-urban population densities. Uniform, comprehensive definitions of "urban," "sub-urban" and "rural" in terms of population density were not found in the literature. However, some indications were uncovered from other geographical regions:

Population size, density and surface area relationships were found for 51 European cities (Stanners and Bordeau, 1995). Among these data, the lowest population density of a denominated city was 485 i/km^2, the smallest population was 12 386 and the smallest city area was 3 km^2; however, about 41% of the cities had population densities ranging from 2 001 to 4 000 i/km^2, the modal city area was in the range 101 to 200 km^2 and about 80% of the cities were greater than 80 km^2 in area.

In a study of land cover and population density in the Baltic Sea drainage basin, a population density of <200 i/km^2 was used to denominate rural and >200 i/km^2 urban (Sweitzer, Langaas and Folke,

1996).

The US Bureau of the Census considers a combination of 50 000 inhabitants and 386 i/km^2 (=1 000 i/mile^2) to be an urban area.

In a study on population growth and agricultural change in Africa, Turner, Hyden and Kates (1993) were able to find "rural" districts of more than 200 i/km^2 uninfluenced in population density by urban concentrations within them. The "rural" districts had densities of up to 416 i/km^2 and a median of 244 i/km^2.

For this study, a population density of 300 i/km^2 was used to indicate the limit at which land would become too expensive for small-scale fish farming. At a nominal grid cell size of 81 km^2, this density corresponds to 24 300 individuals. It was assumed that each grid cell defines the limits of the local market.

Table: Thresholds to indicate potential for farm-gate sales, and the corresponding market size.

Description	Threshold ranges (i/km^2)	Potential market (maximum total inhabitants)
Very Suitable	150–300	24 300
Suitable	25–149	12 069
Moderately Suitable	1–24	1 944
Unsuitable	1 < and > 300	n/a

SOIL AND TERRAIN SUITABILITY FOR FRESHWATER FISH PONDS

A good understanding of soil and its characteristics is one of the most important of the many factors which must be considered for successful freshwater fish culture (Coche and Laughlin, 1985). According to Yoo and Boyd (1993) the common soil properties most relevant to pond construction are slope, texture, organic matter content and sulphide content. The objectives herein were to:

- identify the soil and terrain parameters of the Digital Soil Map of the World (DSMW) (Version 3.5; FAO, 1995) most relevant to fish pond construction and operation, and
- establish thresholds using the class ranges already established

for each parameter in the original development of the FAO-Unesco Soil Map of the World (FAO/Unesco, 1974).

Analytical Approach

Appropriate soil parameters and their corresponding threshold values were taken from tables with soil limitation ratings for excavated fish ponds and for pond embankments, dikes and levees provided by Yoo and Boyd (1993). Index properties of soils for freshwater fish culture were taken from data tabulated by Coche and Laughlin (1985).

The DSMW was used to derive the soil and terrain suitability for fish ponds. This map is based on the FAO/UNESCO Soil Map of the World.

The original scale of the FAO/UNESCO Soil Map of the World is 1:5 000 000. The digital version, on rasters of 5 arc minutes, combines continental-level maps of soil associations and soil groups and derived soil properties as attributes. Included are software routines that interpret the maps in terms of agronomic and environmental parameters such as pH, organic carbon content, C/N ratio, clay mineralogy, soil depth, soil and terrain suitability for specific crop production, soil moisture storage capacity and soil drainage class. The output is given in the form of maps and data files which can be stored for later retrieval.

During the first attempt to classify the soil characteristics as to their suitability for fish ponds, much of Latin America scored low due to stringent thresholds on soil texture. Originally, the optimum class of soil texture was described as "Clayey without swell-shrink, and not organic" and the Suitable to Moderately Suitable class was "Loamy and not organic". The thresholds were relaxed somewhat to "Loamy or clayey without swell-shrink, and not organic".

Table below Soil parameters, suitability ranges and corresponding thresholds for estimating suitability for fish ponds.

Suitability values corresponding to those in Table above were calculated as percentages for each soil mapping unit and grid cell. For the overall evaluation of the soil suitability for ponds, it was assumed that Very Suitable is twice the value of Suitable, and weighting was carried out accordingly.

Soil and terrain property	Very Suitable	Suitable to Moderately Suitable	Unsuitable
Slope %	0 – 2	2 – 8	> 8
Effective soil depth (cm)	> 150	75 – 150	< 75
Gravel and stones %	< 40	40 – 80	> 80
Soil texture	Loamy or clayey without swell-shrink, and not organic		Sandy, or clayey with swell-shrink, or organic
Salinity (dS/m)	< 4	4 – 8	> 8
pH (H_2O; 1:5) (fertility)	7.2 – 8.5	5.5 – 7.2	>8.5 <5.5
Catclays (sulphate toxic, very acid)	Not present	Not present	Present
Gypsum	Not present	Not present	Present

Note: pH measured with a pH meter in the supernatant suspension of a 1:5 soil-water mixture.

WATER LOSS

The main sources of water for levee and contour ponds are precipitation, runoff, pumped or gravity water from perennial water bodies such as streams, rivers, lakes and reservoirs, and pumped groundwater. However, it is difficult to comprehensively account for water availability from all of these sources in a continental-level study of fish farming potential such as this one. For example, perennial streams sufficient to supply a complex of ponds may not appear on digitized maps, and groundwater is not mapped at a continental scale. Therefore, for this study another approach was used, namely the estimation of monthly net water losses from ponds. This approach has the advantage that it indicates the amount of water that would have to be supplied in order to keep ponds full without pre-determining the source of the water. Another advantage is that water loss is estimated in centimetres and therefore no assumption about the area of the pond is necessary.

Water loss from ponds can occur primarily in two ways: evaporation and seepage. Monthly losses via these processes were computed as described below, and then accumulated to estimate annual net water losses.

Evaporation

The objective was to estimate the monthly net loss by evaporation from ponds. The basic investigation of evaporation for this study is described in an internal FAO report (*Estimating water requirements for small ponds in Latin America.* AGLW, 1996).

Two temperature-based formulae were used to estimate evaporation: the Blaney-Criddle formula (Blaney and Criddle, 1950) and the Hargraves formula (Hargraves, Asce and Samani, 1982), of which the latter gave the best results.

The Hargraves approach takes into account air temperature and solar radiation. A coefficient of 1.15 was applied to convert the estimated evaporation into the amount of evaporation that probably occurs in shallow ponds.

Air Temperature

Mean monthly minimum and maximum air temperature data have been collated by the FAO Agrometeorology Group from about 6 500 stations covering Latin America south of the USA, with some stations across the border in the USA. These data were interpolated to obtain a comprehensive air temperature data set at a resolution of 5 arc minutes. The interpolation procedure is briefly described below.

For each month, air temperatures were regressed against a set of variables (linear, quadratic terms) based on elevation and the geographic coordinates, leading to one continental equation accounting for at least 90% of the variance in the temperatures. These equations describe the "climatology" (i.e., the macro-effects). Subsequently, the departures from the "climatology" (residuals) were computed for each station. These corresponded to the local effects, expressed in degrees above or below the "climatology" prediction.

The continent was subdivided into 9 overlapping latitudinal slices of about 20 degrees each. Using the inverse distance weighting programme of Bogaert, Mahau and Beckers (1996), the residuals were then interpolated into a single continental map.

The final steps consisted of adding the interpolated residuals to the computed climatology, and assembly of the slices. The general patterns in the resulting map were comparable to the maps generated by Corbett (1993). However, it is considered that FAO Agrometeorology Group maps offer an improvement in that they are less sensitive to the geographic distribution of the observed points (stations) and better take into account the effects of relief features.

Seepage

According to Yoo and Boyd (1993) seepage can be one of the most important causes of water loss from a pond. They state that seepage rates above 5.08 mm/d (15.4 cm/month, 185 cm/y) are undesirable in levee ponds and that properly constructed ponds often have seepage rates below 93 cm/y and few seep more than 232 cm/y. These authors provide a range of seepage rates and categorize their importance. As a conservative estimate of seepage, a loss of about 100 cm/y was assumed for this study. This corresponds to approximately 8 cm/mo which is within the range of ponds that are categorized as having low seepage rates.

Table: Categories of monthly and annual seepage rates (modified from Yoo and Boyd, 1993)

Category	Rate (cm/mo)	Rate (cm/y)
Low	0–14.7	0–176
Moderate	14.8–30.1	177–362
High	30.2–45.6	363–546
Extreme	>45.6	>546

Water Loss Estimates

Water loss was estimated by balancing inputs from precipitation against losses due to evaporation and seepage for levee ponds:

Water balance

= (Precipitation × 1.1) - (Evapotranspiration × 1.15) - Seepage

The constant 1.1 in the precipitation portion of the above equation accounts for the runoff from the pond's sides that is in excess of the rainfall that falls directly into the pond.

Mean monthly precipitation data on 5-arc-minute grids were employed. These were prepared by Corbett (1993) using a tri-variate thin plate smoothing spline method.

With this approach, areas with two dry seasons could present a problem because the apparent net annual loss would be greater than either of the dry season losses. In order to determine the occurrence of two dry seasons, a set of sample points was defined at approximately 5° × 5°. Only one such area was found, at 25° S, 50 to 55° W, and thus the effect is minor and localized when the entire continent is considered.

The monthly net losses due to evaporation and seepage were accumulated to make a map of annual net water loss.

AGRICULTURAL BY-PRODUCTS AS FEED AND FERTILIZER INPUTS

The presence of agriculture is an important indicator of aquaculture potential in two ways: first, agriculture implies at least a minimum amount of infrastructure for development, such as a road transportation system, a local labour force and villages or towns for essential supplies; and, second, there should be a source of by-products for fish feed or fertilizer.

For small-scale fish farming, agricultural by-products can contribute to higher yields than would be possible from the natural production of the pond. For commercial fish farming, use of by-products from agriculturally produced industrial food processing can reduce feed costs by allowing replacement of part of the formulated feeds otherwise needed.

Unfortunately, there are no comprehensive digital maps of agriculture crop types for Latin America that could be used to infer the availability of by-products as inputs. Therefore, crop potential was used as a surrogate.

Table: Productivity of major land classes in terms of length of the growing period (LGP) and land suitability (from Alexandratos, 1995).

Major Land Class	LGP (days)	Land Suitability
	Dominantly productive land	
AT2 Moist semi-arid	120–179	50% VS+S
AT3 Sub-humid	180–269	50% VS+S
AT4 Humid	270–365	50% VS+S
AT6 Fluvisols and gleysols	Naturally flooded land (NFL)	50% VS+S
	Dominantly marginally productive land	
AT1 Dry semi-arid	75–119	50% VS+S+MS
AT5 Moist semi-arid, sub-humid and humid	120–365	50% MS
AT7 Fluvisols and gleysols mid and humid	NFL	50% MS
	Dominantly unproductive land	
Partly suitable	75–365 or NFL	20–50% VS+S+MS
Mostly not suitable	75–365 or NFL	0–20% VS+S+MS
Not suitable, water, glaciers	n/a	n/a

Crop Potential

Continental views of land with crop production potential are provided by Alexandratos (1995). An evaluation was made of the suitability of land for growing 21 crops under rain-fed conditions and various levels of technology.

Two geo-referenced data sets provided the basic material for the evaluation: the FAO/Unesco Soil Map of the World (1974) and an inventory of climate regimes to characterize length of the growing period (LGP) (i.e., the length of the period of the year when moisture availability and temperature permits crop growth).

The original classification from Alexandratos (1995) was used to identify land with the capacity to produce agricultural by-products as inputs for fish farming.

Thresholds for Inputs

The following assumptions were made in order to re-classify the original rain-fed crop production potential (Table 2.4) in terms of the potential availability of inputs for aquaculture:

- Assumption 1. Longer growing periods potentially provide inputs for aquaculture production over a greater part of the year. Scores for LGP are given in Table 2.6.

Table : Scoring of LGP for potential of inputs for aquaculture.

Score	LGP (days)
4	270 – 365
3	180 – 269
2	120 – 179
1	< 120

In the case of naturally flooded land (AT6, AT7; Table 2.5), an arbitrary LGP score of 2 was assigned in order to account for unavailability of land for agricultural production due to seasonal flooding.

- Assumption 2. Greater productivity per unit area potentially provides greater quantities of inputs per unit area of agriculture production.

Table: Scoring of land productivity for potential aquaculture inputs.

Score	Land Productivity Classification
4	Dominantly productive land
3	Dominantly marginally productive land
2	Dominantly unproductive land
1	Not suitable, water, glaciers

Table: Combination of LGP and productivity scores as inputs for aquaculture.

Original Class	LGP Score	Productivity Score	Scores Combined for Fish Farming
AT2	2	4	8
AT3	3	4	12
AT3	4	4	16
AT6	2	4	8
AT1	3	3	9
AT5	3	3	9
AT7	2	3	6
Partly suitable	2.5	2	5
Mostly not suitable	2.5	2	5
Not suitable	1	1	1

For each 5-arc-minute grid cell, the percentage of each of the original classes was available from the DSMW. These 10 classes were grouped into the four classes shown in Table below. Finally, the percentages of each of the four classes in each grid cell were added to obtain the final score for that grid cell.

MODELS OF SMALL-SCALE AND COMMERCIAL FISH FARMING

The overall objective was to build models that combine the factors described in the preceding sections according to their relative importance for the development and operation of small-scale and commercial fish farming. Obviously, there are likely to be differences in the relative importance of these factors. Also, the importance of the same factors varies between small-scale and commercial fish farming, and according to the intensity of commercial fish farming. For example, the availability of agricultural by-products as feed inputs is relatively important in small-scale fish farming. In contrast, in commercial fish farming, where formulated feeds are used, the availability of such by-products is less important, except where they can be obtained as a cheap supplement to commercial feeds.

Table: Reclassification of LGP-productivity scores into four classes.

New Classes/Scores	Suitability	Score Thresholds
4	Very suitable	> 11
3	Suitable	7 – 10
2	Moderately suitable	3 – 6
1	Unsuitable	< 3

FACTORS FOR SMALL-SCALE AND COMMERCIAL FISH FARMING DEVELOPMENT AND OPERATION

The factors are of two types: those, such as soil suitability for ponds, that are continuous; and constraints, such as reserved areas, that exclude some areas altogether. Constraints are dealt with in the final part of this section.

To summarize earlier sections, the factors considered to be important for the development and operation of small-scale (S) and commercial fish farming (C) and their interpretation for the models are given in the Box.

Urban Market Potential (Market size and proximity)

S: Urban market potential was not used as a factor in estimating small-scale aquaculture potential. It is assumed that all small-scale aquaculture production is sold or bartered locally.

C: Takes into account population size of urban centres together with travel time over roads to the population centres. The maximum one-way travel time allowed from farm to urban market is 6 hours, and the larger the population of the urban centre, the greater the potential market for farmed fish. Implicit is the idea that urban markets are also locations from which farmed fish can be transported by air to other national destinations, or exported.

Farm-gate Sales

This is based on the population density of the grid cell in which the fish farm resides. It is assumed that the higher the population density in close proximity to the farm, the greater the potential for farm-gate sales.

S: Sales or barter of surplus production is an important consideration in small-scale fish farming in parts of Africa, where even modest amounts are sold or bartered (Kapetsky, 1994) and a similar situation appears to exist in Latin America (Martinez-Espinosa, 1994; 1995).

C: For the commercial fish farmer, the high-volume sales at urban markets are much more important than "walk-in" customers who require staff time for the sale of only a few kilos per customer.

Soil and terrain suitability for ponds

Overall soil suitability is important from both an engineering and a productivity viewpoint. Soil characteristics considered include texture, slope, effective depth, pH, salinity, and the presence of catclays and gypsum.

S: Even though moderate limitations can be overcome without undue costs, for the small-scale fish farmer costs and time have to be minimized. Thus it is more important for the small-scale fish farmer to select a site that requires minimum modifications compared to the commercial fish farmer.

C: From the commercial farmer's viewpoint, the capital costs of corrective measures (e.g., lining, compacting, extra dike material) have to be taken into account, together with the increased running costs to replace water losses if corrective measures are only partially undertaken.

Water Loss

Net annual water loss is based on evaporation and seepage.

S: From the small-scale farmer's viewpoint, it is difficult to replace water losses except naturally by rainfall and runoff. Pumping and piping are commercial-level operations to secure water and do not apply to small-scale aquaculture. Therefore, it is important to be in an area where the pond can be kept sufficiently full by rainfall and runoff to enable year-round fish culture.

C: From the commercial fish farmer's viewpoint, water loss can be thought of as running and capital "water costs" (e.g., pumps, pipes, fuel).

On-farm Agricultural by-products

This factor refers to the potential availability of agricultural by-products for use as feed or fertilizer inputs, and is based on soil productivity and LGP.

S: It is important to be able to feed fish (or fertilize ponds) in order to achieve surplus yields that can be bartered or sold.

C: In commercial fish farming, where formulated feeds are used, the availability of agricultural by-products is less important, except where these can be obtained as a cheap supplement to commercial feeds.

WEIGHTING PROCEDURE

The procedure used to build the small-scale and commercial fish farming models is multiple-criteria, single-objective decision making. It is summarized by Eastman *et al.* (1993) and by Eastman (1995), and therefore described only briefly here. The development of weights is based on pair-wise comparisons. The comparisons concern the relative importance of the two criteria involved in determining suitability for the stated objective. The ratings can be on a 9-point continuous scale as in Table below; but, where it is felt that there is too much contrast between adjacent comparisons, additional comparisons can be added, such as "1/6" between 1/7 and 1/5.

PROCEDURE FOR BUILDING THE COMMERCIAL MODEL

The model for commercial fish farming was based on the multiple-comparison method outlined above. Aquaculture experts from the FAO

Inland Water Resources and Aquaculture Staff and a former FAO field staff member participated.

Table: Expertise and international experience of FAO staff participating in the couple comparisons for the commercial model

Expert	Aquaculture Expertise	International Working Experience
1	General, projects, research	Latin America, Global
2	Small-scale rural development	Latin America
3	Small-scale farming systems, extension	Africa
4	Nutrition	Latin America, Global
5	General, research	Pacific, Middle East, Global

Establishment of values for the couple comparison was accompanied by much discussion based on differing views and perceptions of the factors and their relative importance. Careful definition of the factors was required.

Table: Relative importance of commercial fish farming factors in couple comparisons (row versus column; e.g., water cost is strongly less important than market size and proximity).

	Market size & proximity	Water loss	Soil suitability	Agricultural by-products	Farm-gate sales
Market size & proximity	1				
Water loss	1/5	1			
Soil suitability	1/7	1/5	1		
Agricultural by-products	1/9	1/9	1/7	1	
Farm-gate sales	1/9	1/7	1/9	3	1

After the comparisons were made, the IDRISI Weight module (Eastman, 1995) was used to calculate the eigenvectors and to establish the consistency ratio of the results. The results are shown in the "Experts" column in Table below.

Table: Relative importance of five factors for commercial fish farming development and operation.

Factors	Experts	Author
Market size and proximity	55%	49%
Water loss	25%	27%
Soil suitability	13%	13%
Potential for agricultural by-products	4%	5%
Farm-gate sales	3%	6%

The consistency ratio (0.2) of the expert group was unsatisfactory. However, the group's results were very similar to those generated by the first author using the same method well prior to the session with the experts, and the consistency ratio obtained by the first author was at an acceptable level, 0.07, so his model was used.

COMBINATION OF THE FACTORS FOR THE COMMERCIAL MODEL

The individual factor maps were weighted according to their relative importance as shown in the table above. Because the grid cell of each factor map had already been classified as VS, S, MS or US, each cell was assigned a score obtained as the product of the weight of the factor and the suitability of the particular grid cell pertaining to that factor. These scores were then added for all factor maps. Finally, thresholds were established to arrive at a map of the overall suitability for commercial fish farming in four classes: VS, S, MS and US.

PROCEDURE TO BUILD THE SMALL-SCALE AQUACULTURE MODEL

The small-scale model was established in a similar way to that of the commercial model. However, the role of the group of experts was limited to a discussion that clarified the factors and their relative importance. The multiple criteria comparisons were undertaken by Expert 3 and the first author working together. The consistency ratio,

0.07, was acceptable. The results of the comparison, after calculation by the IDRISI Weight module, are shown in Table below.

Table: Relative importance of factors for the development and operation of small-scale fish farming.

Factors	Relative importance
Water Loss	51%
Potential for agricultural by-products	19%
Soil suitability	16%
Farm-gate sales	14%

The factors were combined to obtain a small-scle fish farming suitability map in the same way that the commercial map was constructed (as discussed above).

FISH FARMING YIELD POTENTIAL OF FOUR SPECIES

Fish growth models are useful tools for analysing factors that affect fish performance under culture conditions, and as management tools for fish production (Cuenco, Stickney and Grant, 1985). The use of such models in regional-scale planning applications (e.g., assessment of the production potential for different aquaculture species by the use of GIS tools) is not yet common, presumably because available models have either not been adapted for the wide variety of species needed to reflect aquaculture potential over large areas that encompass wide variations in climatic conditions, or they have not been packaged in a form that permits integration with GIS. Another possible reason is that such models can be used only when the relevant spatial and digital data are available to serve as inputs. Thus, the basic task here was twofold: to provide a synoptic water temperature data set for Latin America that could be used as one of the inputs of a growth model, and to modify an existing model to be used with GIS to provide yield outputs for four species.

WATER TEMPERATURE MODELLING

Validation of this model has been accomplished for different geographical locations by the use of daily weather and site-specific input data recorded in the tropical aquaculture database maintained by PD/A CRSP (Nath, 1996).

However, daily weather data required for the water temperature model are not easily available or accessible for large geographical regions such as Latin America. Moreover, manipulating and storing daily weather data for such a large grid are not trivial tasks. Even if capabilities for such operations are available, it is unlikely that there would be significant advantages of using them for strategic planning applications such as continental-scale analysis of aquaculture potential. An alternative approach taken in the current study was to procure from the FAO Agrometeorology Group gridded data sets of the monthly values of air temperatures required for the water temperature model. Unfortunately, due to problems encountered during interpolations of recorded weather data, gridded data sets of the other weather variables required as input to the water temperature data sets were not available. Air temperature data were interpolated to obtain daily values for use in the water temperature model; other input variables were predicted by the use of a simple weather generator developed by Fritz, Meredith and Middleton, (1980). The analysis assumed that all the simulated ponds have a constant pond area of 2000 m^2 and a depth of 1.2 m. Pond volume was consequently also assumed to be constant (i.e., 2 400 m^3).

One-year simulations were conducted for each cell within the grid for Latin America using a one-day time step and the resulting daily water temperature values averaged for each month. The latter set of values was then used to generate maps of monthly mean water temperature. Because the model has been adequately tested only for conditions where the water temperature exceeds 0°C, a condition was inserted during map creation to plot the mean monthly air temperature instead of water temperatures < 0°C.

FISH GROWTH MODELLING

For use in the current study, an existing fish growth model (Bolte, Nath and Ernst, 1995) was modified. It is briefly described, with calibration and validation results.

One-year simulations were conducted with the fish growth model for all four species. Daily water temperatures required for the growth model were obtained by linear interpolation of monthly means predicted by the water temperature model above. It was necessary to make minor modifications to the growth model in order to generate crops/y output. For example, it is difficult to specify exact stocking and harvest times for simulation runs relevant to a large geographical region because the time period required to reach the desired harvest size varies between locations. Further, depending on temperature preferences of different species, fish culture may be possible only during a certain period of the year, which is difficult to predict *a priori*. It was therefore assumed that a fish lot (population) would be stocked when water temperatures were favourable (arbitrarily defined as a 15-day period for which temperatures are within the range for which growth of the particular fish species is possible. The desired harvest weight for each species was also specified before a simulation run commenced.

During the simulations, the fish population was harvested if any of the following conditions were encountered:

- the specified harvest weight (i.e., a full crop) was reached before one full year.
- the simulation duration of one year was reached without registration of one or more crops, or
- water temperatures were unfavourable for growth.

For the first case, if water temperatures continued to be suitable for the selected fish species, it was assumed that an additional fish lot would be stocked and harvested according to the above conditions. For the second and third cases, the crops/y output was expressed as a decimal fraction of the current fish weight relative to the harvest weight (i.e., a partial crop was reported).

COMMERCIAL FARMING

In order to explore a range of commercial aquaculture possibilities by the use of the growth model, simulations were conducted for all four fish species, assuming two different feeding levels and two harvest weights. Natural food availability was assumed to be negligible. Feed application rates corresponding to 50% and 75% satiation were assumed to represent commercial aquaculture operations with low and high feeding rates, respectively. The primary effect of higher feeding rates as realized by the model is to enable more rapid fish growth, thereby enabling the desired target size to be reached earlier. The advantage of using percent satiation feeding levels instead of feeding rates on a percent body weight basis is that the former approach takes into account variations in the factors affecting fish appetite (see also Annex 3) during calculation of feed requirements. Within each of the percent satiation feeding levels, two harvest weights (small and large) were also established to enable consideration of market preferences for each fish species.

The stocking weight for tilapia, tambaqui and pacu was assumed to be 50g for all the scenarios; carp were stocked at 100g. A survival rate of 80% was assumed for all the simulation runs.

Species	Stocking Density (fish/m²)	Harvest Size (g)	Gross Yield (kg/ha/yr)	Net Yield (kg/ha/yr)
Tilapia	3.0	300	7 200	5 700
	1.5	600	7 200	6 450
Tambaqui	1.5	600	7 200	6 450
	0.9	1 000	7 200	6 750
Pacu	1.5	600	7 200	6 450
	0.9	1 000	7 200	6 750
Carp	1.25	600	6 000	4 750
	0.5	1 500	6 000	5 500

To achieve the overall objectives of this study and for easy interpretation of the results, it was necessary to aggregate crops/y outputs from the simulation runs into four classes. However, specification of rigid classes that pre-judge the value of the output without accompanying production cost and marketing data would not be appropriate. Furthermore, differences in model output were expected depending on the particular species, harvest size and feeding levels. To avoid these problems, output for each simulation scenario was divided into equal quarters of the range of crops/y. They are designated as 1st quarter (highest crops/y) (1^{st}Q), 2nd quarter (2^{nd}Q), etc.

SMALL-SCALE FARMING

It is difficult to precisely define small-scale, subsistence-level aquaculture operations because of the wide variety of materials used as inputs to these systems and variation in fish sizes that are harvested. In general, however, such systems are characterized by low-intensity management and smaller sizes of fish at harvest. Analysis of the potential for small-scale farming was limited to tilapia and carp only, because these two species effectively utilize natural food resources in ponds, as opposed to tambaqui and pacu which perform well primarily in ponds that receive artificial feed of relatively high quality.

Natural food availability was modelled as a function of fish biomass (Bolte, Nath and Ernst, 1995), an approach that requires definition of the critical standing crop (in kg/ha) or critical fish biomass (CFB; in kg/m^3). For tilapia ponds that are not heavily fertilized or fed, a CFB of about 0.075 kg/m^3 (equivalent to a fish biomass of 750 kg/ha) appears to be reasonable (Bolte, Nath and Ernst, 1995). This value also was assumed for carp ponds in the current study.

Species	Stocking density (fish/m^2)	Stocking size (g)	Harvest size (g)	Gross Yield (kg/ha/y)	Net Yield (kg/ha/y)
Tilapia	2	25	150	2 400	1 900
Carp	1	50	350	2 800	2 300

Model output as crops/y for tilapia and carp under small-scale simulation conditions was also divided into equal quarters (as described in the section on commercial farming above).

COMBINATION OF ESTIMATES OF POTENTIAL FROM COMMERCIAL AND SMALL-SCALE MODELS WITH THE SPECIES YIELD MODELS

As shown above, the factors important for the development and operation of fish farming have been brought together in the form of commercial and small-scale model suitability maps, each with four classes ranging from VS to US. Similarly, the crops/y results from the growth models have been aggregated into four equal-length quarters for all four species within each commercial and small-scale fish farming scenario. In the final analytical step, these outputs were combined to make maps that showed the coincidence of each class of suitability with each quarter of crops/y output. For clarity in presentation of the maps, only VS, S and MS areas are shown in combination together with only the 1st, 2nd and 3rd quarters of crops/y yields.

Constraints

The constraints included in this study were large inland water bodies and protected areas. Data for the latter constraint were obtained from the International Union for Conservation of Nature and Natural Resources (IUCN, 1992). Unfortunately, not all of the protected areas of Latin America are available in the form of digital maps. Thus, it is unavoidable that some protected areas may be shown in this study as having potential for fish farming. These constraints were brought into the analyses at the fourth step of the study, at the point where the species crops/y outputs were combined with the commercial and small-scale model outputs.

Verification

The objective was to test the predictions of aquaculture potential made by use of the commercial model in combination with the yield model in terms of suitability of the grid cell "sites" and their yield in crops/y. The actual locations of operational farms were used as the arguments. The verification study was confined to one country, Colombia, due to financial constraints. Nevertheless, Colombia is an important producer of cultured fish from inland waters.

Table: Production from inland aquaculture in Colombia in 1993 (Fishery Information Data and Statistics Service, 1995)

Species	Production (tonnes)
Oreochromis niloticus	11 046
Piaractus brachypomum	3 500
Oncorhynchus mykiss	2 028
Cyprinus carpio	60
Macrobrachium rosenbergii	10
Total	16 644

Also, the mix of species indicates that Colombia is heterogeneous with regard to conditions for aquaculture development and for this reason, too, it is a good testing ground.

Farm characteristics were obtained by interview and locations by use of a global positioning system (GPS) instrument for 34 commercial farms among the eight departmentos (provinces) important for fish farming in the N, NW and central W portions of the country. The data were collected and compiled by personnel of the Instituto National de Pesca and Acuicultura from April to June, 1996, working under contract to FAO.

Results

For convenience and according to the focus of the reader, the results are apportioned into continental and country-level views. Also, the results are partitioned from another perspective: those for the culture system and those for the species cultured.

That is, factors important for the development and operation of fish farming at commercial and small-scale levels are evaluated separately from the crops/y yield potential for each species based on water temperature and feeding rate × harvest weight combinations. This separate evaluation was made because one may want to examine the estimates of the quality of the land for fish farming in ponds irrespective of the species used. For the same reason, yield potential for each species is considered separately. Finally, the factors and the species yields are considered together.

CONTINENTAL OVERVIEW OF FACTORS FOR FISH FARMING DEVELOPMENT AND OPERATION

About 56% of the continent is within reasonable proximity of large urban centres that could be important markets for commercial fish farming development, or as transit points for export. This is significant for commercial farming because urban market potential was deemed the most important factor among the five considered. Farm-gate sales and soils for fish ponds are the most spatially limiting of the factors considered here, with only 1.1% and 5.5% respectively of the area of

Latin America being categorized as VS[1]. The areas that are favoured for farm-gate sales are parts of Central America and the coastal fringes of NW, NE and E South America, where populations are relatively high. Soil and terrain suitability for ponds has a patchy distribution, with Central America and southern South America favoured.

In contrast, about 22% of the land area in Latin America is VS for agricultural by-products and 19% VS for water loss. Central America is at an advantage for the former, but much of the western fringe of South America is not. Water loss is relatively high in NW Central America and in the W, SW and NE of South America.

Table: Suitability of factors for fish farming development and operation as a percentage of the surface area in continental Latin America.

Factor	VS	S	MS	Total
Urban market demand	56.2	17.6	8.4	82.3
Farm-gate sales	1.1	17.1	70.7	88.9
Soils	5.5	36.3	30.0	71.9
Agricultural by-products	22.3	38.5	27.7	88.5
Water loss	19.4	43.0	19.6	82.0

In all, from about 72% to 89% of the continent ranges from MS to VS among the five factors considered.

1. The categories used and their abbreviations are: Very Suitable (VS); Suitable (S); Moderately Suitable (MS); and Unsuitable (US). The yield ranges are divided into four equal quarter classes: 1stQ; 2ndQ; etc.

CONTINENTAL OVERVIEW OF COMMERCIAL AND SMALL-SCALE FARMING MODELS

When the five factors discussed above are combined to make the commercial model, the result is that about 20% of Latin America scores VS, about 38% S and about 22% MS. For the small-scale model, about 31% of the continent scores VS, nearly 44% S and 15% MS. Central America, NW and SE South America are areas where both commercial and small-scale farming are favoured.

COUNTRY-LEVEL VIEWS OF FACTORS

Urban Market Potential

The effect of proximity to large urban centres can be clearly seen as circular areas on the map. Five countries, Belize, Costa Rica, El Salvador, Guatemala and Uruguay, have 100% or nearly 100% of their national areas scoring VS for this factor. Eleven other countries have from 50% to nearly 90% of their areas in this category. Only Suriname has a low overall score, with just over 50% of its area that is at least MS.

Potential for Farm-gate Sales

Only one country, El Salvador, has a large area (about 60%) that is VS for farm-gate sales, whereas most of the other countries have 5% or less of their area in this category. Only one country, Honduras, has more than half of its area that is suitable for farm-gate sales. There are six additional countries that have more than 25% of their national area classed as suitable.

Soil Suitability for Ponds

Uruguay and Argentina respectively have about 25 and 20% of their area that scores VS. There are eleven countries that have 50% or more of their soil areas that score VS or S together. French Guiana and Guyana are notably relatively lacking in good quality soils for ponds.

Potential Agricultural By-products

Three countries, El Salvador, Suriname and Uruguay, have nearly one-half of their national areas that score VS for potential agricultural by-products, but most countries have much of their area that scores either S or MS for this factor. Chile is the only country with less than one-half of its national area that is not at least MS in potential for agricultural by-products.

WATER LOSS

Seven countries score VS for 50% or more of their areas and eight others have from 25 to 50% of their area that is VS. Mexico and Argentina stand out as countries with relatively small areas that score VS or S for water loss, suggesting that water requirements could be a relatively serious constraint there.

COUNTRY-LEVEL VIEWS OF COMMERCIAL AND SMALL-SCALE MODELS

Commercial Farming

Generally, the countries of Central America, NW and SE South America are favoured in suitability for commercial farming. Three countries, Costa Rica, Belize and Guatemala, have more than 90% of their area classed VS for commercial farming. Six other countries have more than 50% of their national area that scores VS. For five of them, all, or most of the remaining area, is classed S. French Guiana and Suriname have relatively small areas that are VS or S, although these together have a combined total of more than 30%. Brazil has the least area that rates MS or better, although the combined total exceeds 60%.

SMALL-SCALE FARMING

Two countries, French Guiana and Suriname, have more than 90% of their area that is VS for small-scale farming and there are eight additional countries that have more than 50% of their area in the VS category. Additionally, for the same countries the remaining area is all S, or nearly so. Mexico is least favoured; although even there about 60% of the country is MS or better for small-scale farming, of which about 13% is VS and 25% S.

Fish Yields

Yields in terms of crops/y are the key output from the growth model for the four different fish species. This output is presented first from a continental viewpoint and then from a country viewpoint. The continental country-level results are expressed in terms of relative surface area. For simplicity, crops/y results have been separated into quarter parts of the ranges that have been attained with each feeding rate × harvest weight combination. Again, for simplicity, only the first three quarters are presented as histograms, but all four quarters are mapped.

Continental View

Relatively large areas of Latin America are suitable for the commercial farming of the four species considered here. For example, areas in which 1stQ (i.e., highest) crops/y are attainable range from a high of 74% of Latin America for carp to a low of about 36% for Nile tilapia, except for one result for the latter species that is only 11%. Results for individual species range from 67 to 74% for carp, 11 to 45% for Nile tilapia, 57 to 67% for tambaqui and 49 to 61% for pacu.

Significantly, areas corresponding to 2ndQ and 3rdQ crops/y are small in comparison to 1stQ areas, except for the 50% satiation feeding rate of Nile tilapia. Therefore, most of the area that is apt for farming of these four species is capable of producing relatively high numbers of crops/y in each feeding rate × harvest weight combination.

Within individual species results, there is relatively little difference in the 1stQ surface areas that result from different feeding rate - harvest size combinations. Rather, it is the numbers of crops/y that vary markedly when the different regimes are simulated. As would be expected, it is the combination of low feeding rate and high harvest weight that tends to produce the least crops/y and conversely, the high feeding rate - low harvest weight that results in the highest number of crops/y. In general, feeding at the high rate (75%) provides a good combination of relatively large surface areas and high number of crops/y. Feeding at 50% and harvesting at the high weight clearly produces the lowest crops/y along with the least surface area, except for tambaqui, for which the lowest 1stQ area is obtained with 50% feeding and the low harvest size.

For small-scale farming, the results vary considerably between the two species considered. For Nile tilapia harvested at 150g, a relatively small area of Latin America (36%) can yield 1.3 to 1.7 crops/y and an additional 13% of the surface area can produce 0.9 to 1.3 crops/y. In contrast, 72% of the continent can produce 1.4 to 1.8 crops/y of carp harvested at 350g, and another 12% of the area can produce 0.9 to 1.4 crops/y.

COUNTRY-LEVEL VIEWS OF YIELD POTENTIAL BY SPECIES: - COMMERCIAL FARMING

In this section we generalize the spatial distribution of species potential on a country by country basis. Each feeding rate - harvest weight combination is illustrated by a histogram. However, as indicated above, among feeding rate × harvest weight combinations within species there is relatively little surface area difference, with one exception in the case of the Nile tilapia. Further, it is difficult to discriminate such small area differences on the continental maps reduced to A4 size for this publication. For these reasons and also for economy in publication costs, only one map is used to illustrate the spatial distribution of each species. As has been shown above, it is the combination of feeding at 75% satiation with the low harvest weight that provides the highest numbers of crops/y along with relatively large surface areas. Therefore, this combination has been selected for the map illustration.

Tambaqui

The spatial pattern for tambaqui is similar among the feeding rate - harvest weight combinations: 1stQ crops/y are possible in S Mexico and in all or nearly all of the national areas of the Central American countries south of Mexico, throughout the N and NW of South America apart from the Andes, over much of Brazil (apart from the SE) and parts of Bolivia, Paraguay and Peru. It is all but N Argentina, Chile, N Central Mexico and Uruguay that are disadvantaged.

Pacu

The spatial pattern for pacu is similar to that of tambaqui; however, it is more restrictive in that 1stQ crops/y areas are less and 2ndQ crops/y areas are larger.

Nile Tilapia

The spatial distribution of commercial culture opportunities for the Nile tilapia is markedly more restrictive than for the other species; however, the same group of Central American and N and NW South American countries maintain 1stQ crops/y potential, but much of the relatively high yield potential is lost in the marginal countries.

Carp

As would be expected for a species with a relatively wide temperature range for growth, the spatial distribution for carp culture is greater compared to the other species. For example, 1stQ and 2ndQ crops/y ranges can be realized over much of Mexico and in N. Argentina, whereas only small areas of these regions are apt for the other three species.

COUNTRY-LEVEL VIEWS OF YIELD POTENTIAL BY SPECIES: - SMALL-SCALE FARMING

Carp and Nile tilapia contrast greatly in the spatial distribution of their small-scale-culture potential. Nile tilapia is apt for small-scale farming in the same countries as carp and 1stQ and 2ndQ crops/y ranges are similar to those of the carp, but Nile tilapia potential extends over a smaller area. Nevertheless, 1stQ crops/y (1.3 to 1.7) can be obtained in 50% or more of 11 countries, and only Uruguay and Chile offer no possibilities for yields in the 1stQ and 2ndQ ranges. Opportunities for small-scale farming of carp are extensive. 1stQ yields of from 1.4 to 1.8 crops/y can be attained in more than 50% of the areas of all but three of the countries and it is only Chile that is quite disadvantaged.

POTENTIAL WITH SPECIES YIELD AND COMMERCIAL AND SMALL-SCALE MODELS COMBINED

In this section we examine species yield potential as influenced by the five factors that affect the development and operation of fish farming in ponds after they have been brought together in the commercial model and four of the same factors which have been combined in the small-scale model. In order to draw attention to those areas of Latin America where conditions are most favourable for inland fish farming, as well as for simplicity and clarity, we discuss the spatial aspects of only the highest ranking combinations: areas that score VS or S for the commercial model and areas that are apt for fish crops in the 1stQ and 2ndQ ranges of crops/y. However, the maps do indicate all of the combinations of model outcomes from VS to MS, and crops/y outcomes for the 1stQ, 2ndQ and 3rdQ ranges.

Continental View - Commercial Farming

Even with the constraints imposed by meeting the factor requirements of the commercial model along with those of favourable water temperatures, there are large areas of Latin America that are very apt for commercial fish farming of the four species under consideration herein. Area-wise, the pattern shows very little difference among feeding rate × harvest weight combinations within species. Among species, areas that score S-2ndQ to VS-1stQ range from a low of 19 to 25% for the Nile tilapia, from 25 to 35% for pacu, from 33 to 36% for tambaqui, and up to 40 to 44% for carp. This suggests that, in terms of areas suitable, opportunities for carp farming are considerably better than for Nile tilapia, while tambaqui and pacu are intermediate.

As would be expected, the pattern established earlier - suggesting that the high feeding rate combined with the low harvest weight provides the largest area, and also the highest number of crops/y - also holds when the commercial model is brought in. However, as noted above, the range of differences in areas among the various feeding rate × harvest weight combinations is not great.

Continental View - Small-scale Farming

Within the small-scale model, 60% of continental Latin America is very apt for farming of carp and will provide 0.9 to 1.8 crops/y. About 24% of that area scores VS and could provide from 1.4 to 1.8 crops/y.

In contrast, the area that is very apt for Nile tilapia small-scale farming and which will provide 0.9 to 1.7 crops/y is about 38%. About 16% of this area scores VS and could provide 1.3 to 1.7 crops/y.

Country-level View - Commercial Farming

As was pointed out above, there is little spatial difference from one feeding rate × harvest weight strategy to another within a species. Thus, the spatial distribution can be adequately illustrated by a single map for each species. We have chosen the high feeding rate × low harvest weight combination map for each species because, as already established, this strategy results in the highest crops/y over the largest expanses of area. However, histograms of the spatial distribution pertaining to each feeding rate × harvest weight combination are presented.

Tambaqui

The countries most favoured for the commercial farming of this species are all in Central America. There are six countries that have more than 50% of their area that scores VS-1stQ (1.1 to 1.4 crops/y) for the

75%-600g combination. Eleven additional countries have at least 30% of their areas that provide a minimum of S-2ndQ possibilities.

Pacu

The results for this species are nearly identical to those for tambaqui in terms of spatial distribution and rank order of the countries in area, but for the 75%-600g combination, higher yields, of 1.5 to 2 crops/y, are included in the 1stQ range of crops/y.

Carp

As with tambaqui and pacu, Central American countries score high, with more than 40% of the national area being VS-1stQ (1.7 to 2.3 crops/y) for the 75%-600g combination. However, other countries possess relatively large areas that provide at least S-2ndQ possibilities, among them Mexico and Uruguay.

Nile Tilapia

Although the Central American countries dominate, the areas that score highest are considerably less than for the other species. There are only three countries, Belize, Costa Rica and El Salvador, that have more than 50% of their area classed VS-1stQ (1.8 to 2.4 crops/y in the 75%-300g combination) and only one other, Panama, has in excess of 40% of its area that falls in the same category. Nevertheless, there are 11 additional countries that equal or exceed S-2ndQ possibilities over more than 30% of their national areas.

Country-level View-small-scale Farming

Unlike commercial-level farming, two of the northern South American countries - French Guiana and Suriname - score highest both for Nile tilapia and carp farming, and the ranks for the remaining countries are quite similar between the two species.

Eight countries have more than 50% of their area that is VS-1stQ (1.4 to 1.8 crops/y) for carp harvested at 350g, whereas there are only one-half that number for Nile tilapia harvested at 150 g in the same category (1.3 to 1.7 crops/y). In all, there are 20 countries that meet or exceed S-2ndQ requirements for over 20% or more of their national areas for carp and 18 for Nile tilapia. Eight of these countries have more than 80% of their rated areas in this category for carp and 5 countries for Nile tilapia. Suitability for commercial farming and potential yield (crops/y) of tambaqui fed at 75% satiation and harvested at 600 g (mc750600)

Verification

This section compares the predictions of suitability for commercial fish farming and predicted yield potential with observations from 34 commercial fish farms in Colombia.

All of the 34 farms raised red tilapia (*tilapia roja*) alone or predominately. Nearly all farms were using semi-intensive or intensive methods.

The weights at harvest of red tilapia among 32 commercial farms averaged 405 g and the median was 378 g.

Thus, considering species cultured, intensity of farming and harvest weight, the most appropriate grid cell "site" comparison with farm locations corresponded to the predicted potential for commercial farming of Nile tilapia fed at 75% satiation and harvested at 300 g.

The outcome of this comparison was that 20 of the farms were in grid cells predicted to be VS by the commercial model, of which 13 farms were in cells predicted to yield crops in the 1stQ range, four in the 2ndQ range and three in the 3rdQ range. Fourteen farms were in grid cells either US from the viewpoint of the commercial model, or with crops/y falling into the 4thQ range.

Table: Farms raising red tilapia, cachama and other species among those surveyed.

No. of farms	Species mix
10	Tilapia roja alone
20	Tilapia roja with cachama alone, or with others
3	Tilapia roja with others, but not cachama
1	No data
34	Total

Table: Intensity of farming among 32 farms

No. of farms	Intensity of farming
1	Extensive
23	Semi-intensive
8	Intensive
2	No data
34	Total

Table: Weights at harvest of red tilapia from 32 commercial farms in Colombia

Weight at harvest of red tilapia (g)	Number of farms
201–250	4
251–300	5
301–350	4
151–200	1
351–400	5
401–450	2
451–500	7
501–550	1
551–600	2
> 600	1
No data	2

For the purposes of verification, it is most important to know why the predictions of commercial suitability and yield potential did not correspond to the locations of the active commercial fish farms so that future versions of the models can be refined. Therefore, the 14 farms locations that were US for Nile tilapia were the focus of subsequent investigations.

In order to further pin down the causes for the disparity at the 14 US grid cell sites, the 34 farm locations were compared with predictions of suitability made by the commercial model alone. The outcome was that all the sites that were US were, in fact, VS for the commercial model. Therefore, the apparent problem at the commercial farm locations was not with any of the factors upon which the commercial model is based, but with the growth model that predicts yield in crops/y. Because the most important input factor of the yield model is the mean monthly temperature regime, possible temperature anomalies at the 14 farm locations were investigated.

Table: Predicted suitability of grid cell sites containing 34 fish farms for commercial farming of Nile tilapia fed at 75% satiation and harvested at 300 g

Score for the commercial model	Score for yield in crops/y (quarter ranges)	Number of farms
VS	1st Q	13
VS	2nd Q	4
VS	3rd Q	3
US (i.e., US for the commercial model, or crops/y in 4th Q)		14

In order to test the hypothesis of temperature anomalies, the 34 farm locations were compared with the predicted for carp fed at 75% satiation and harvested at 600 g. It was reasoned that, if the problem is indeed one of temperature, then more farm locations should be suitable for carp than for Nile tilapia because the former species is more tolerant to lower temperatures than the latter.

The outcome of this comparison was that all but one site was in the 1stQ range of crops/y for carp and the remaining site was in the 2ndQ range. This suggests that, indeed, the problem is related to temperature at the 14 fish farm locations US for Nile tilapia. With regard to the predictions of tilapia yields, the problem can have three sources:

- predicted water temperatures are lower than actual temperatures,
- the model is not adjusted to reflect actual farming conditions with respect to water temperature, or
- the mainly red tilapia strains grown in Colombia are more cold

tolerant than those upon which the model is based.

Predicted vs Actual Temperatures

One way that predicted temperatures for grid cells can be lower than actual temperatures at farm sites is related to the terrain and grid cell size in relation to farm size. The nominal dimensions of a grid cell are 9 × 9 km, or 81 km^2 whereas the farms are a fraction of this area, as half of the 34 commercial farms were less than 5 ha in extent.

Table: Commercial farm sizes among a sample of 34 in 8 **departmentos** ***in Colombia***

Size Interval (ha)	Number of Farms
1–5	17
6–10	7
11–15	3
16–20	3
21–25	1
26–30	1
> 30	2
Total	34

The water temperature predicted for a grid cell characterises its entire area based on air temperature which, in turn, is partly based on altitude. Another way of expressing the same idea is that farms inhabit a microclimate compared to the "macroclimate" of the grid cell in which they are located. This hypothesis was tested by examining the terrain in the vicinity of the fish farms that were US for Nile tilapia, mainly using 1:25 000 scale maps, or else 1:100 000 maps.

As would be expected, in the hilly and mountainous terrain of W Colombia, nearly all of farms are located on relatively flat land in valleys. To generalize, of those that were US, many are in the range of 600–1 000 m and there are altitudes of from 1 300 m up to 3 000 m within relatively short distances of about 4 to 10 km from them. Two of the farms that were US are at about 1 350 and 1 500 m. Thus, it is highly likely that, because of the surrounding terrain, the estimated temperatures of the grid cells in which the farms fall are less than the actual temperatures at the farm sites.

In order to test the general correspondence between predicted and observed water temperatures, the grids of mean monthly temperature were queried for the grid cells with temperature observations and also grid cells encompassing the locations of the three fish farms (labelled F1, F2 and F3 respectively) that were both US for tilapia and at the highest altitudes. The water temperatures reported for Farms 1, 2 and 3 were in the ranges of 21–30°, 20–27° and 23–25°C, respectively. In comparison, predicted temperatures for the same locations ranged from 22.2–23.4°, 19.8–21.4° and 22.4–23.7°C. These data suggest a tendency towards both lower values and a narrower range for predicted temperatures, which may in part account for poor predictions of Nile tilapia yields in the grid cells where active farms exist.

Another aspect of this same problem is that predicted water temperatures were based only on actual air temperature inputs; other model inputs were from our weather generator, which gives "reasonable" guesses at best.

Different Strains of Tilapia being Grown on the Farms

It is possible that the mainly red tilapia strains used at the locations are either more cold-tolerant than "typical Nile tilapia" or have adapted to ambient conditions on the farms. For example, cold-adapted strains of Nile tilapia from Egypt were successfully grown at high altitudes and low temperatures in Rwanda (Nathanael and Moehl, 1989).

Model Parameters

The lower temperature threshold in the growth model is about 18.6°C for Nile tilapia. Model parameters were based on growth data from Honduras. This lower threshold may be a little too high because Nile tilapia is capable of survival, if not growth, at temperatures as low as about 16°C. However, we settled on a lower threshold of 18.6°C because it provided reasonable predictions of yields in locations as diverse as El Carao, Honduras, and Bangkok, Thailand, during the model validation phase.

DISCUSSION

Potential for Inland Fish Farming in Latin America

The results are quite positive for the development of inland fish farming in Latin America. They show that both commercial and small-scale fish farming is possible over vast areas without serious constraints, either from the lack of suitability of basic factors that are important for development and operation of fish farms, or from constraints of temperature on fish growth. One of the most significant findings of this study is that, from a commercial fish farming viewpoint, combinations of relatively high population urban centres and nearby road infrastructure place large areas that are suitable for fish farming within easy reach of urban markets. Another significant finding is that large areas of Latin America are suitable for the farming of a variety of species. The numbers of crops per year can be maximized by a combination of relatively high feeding rates and harvesting at moderate weights.

From a country-by-country viewpoint, the results are also generally positive. For small-scale farming of Nile tilapia, 19 of the 21 continental countries have 6 to 100% of their national areas that rate at least suitable and from which at least 0.9 crops/year can be obtained. The corresponding results for small-scale farming of carp are that 20 of 21 countries have some areas that are at least suitable and these range from 24 to 99% of national surface areas. From these areas, at least 0.9 crops/y are possible. Commercial fish farming of Nile tilapia with the 75% feeding rate × low harvest weight combination, is possible in 18 of 21 of the countries. The surface area of these countries that rates at least suitable ranges from 14 to 95%, and at least 1.2 crops/y can be obtained. The corresponding results for carp indicate that fish farming is possible in all 21 countries, with 4–99% of their national surface areas rated at least as suitable. Further, at least 1.2 crops/y can be raised. Of the 21 countries, 19 and 18 can support tambaqui and pacu culture respectively, with the corresponding surface areas ranging from 8 to 99% and 18 to 98%. At least 0.7 and 1 crops/y are possible for tambaqui and pacu, respectively.

Latin America has a larger relative area than Africa with potential for inland fish farming. That is, about 31% of Africa encompasses potential for small-scale farming of Nile tilapia and about 13% encompasses potential for commercial farming of the same species from which 1 to 2 crops/y can be obtained (Kapetsky, 1994). In comparison, 38% to 60% of Latin America scores suitable to very suitable for small-scale farming of Nile tilapia and carp, respectively. In the same areas, 0.9 to 1.8 crops/y can be realized. For the same two species, 19% to 44% of Latin America scores suitable to very suitable for commercial farming, and 1.2 to 2.4 crops/y can be realized on the same areas. In absolute terms, the situation is reversed. Continental Latin America is about 60% of the area of Africa. In these terms, Latin America has about T! as much area as Africa that is appropriate for farming of Nile tilapia at the small-scale level and about 80% as much of Africa that is suited for commercial farming.

Interpretation of Factors and Model Results

The estimates of area with potential for fish farming development (or lack of it) are influenced by many factors. Among them are the use of surrogate factors, the selection of thresholds that describe various classes of suitability, and the use of nominal area estimates.

Surrogate Factors

Surrogate factors had to be used where data on actual factors were not available. For example, in order to estimate the availability of agricultural by-products as feed and fertilizer inputs for fish farming, it was necessary to use estimates of land suitability for agriculture because there are no comprehensive maps that show the actual spatial distribution of agriculture at a resolution that would have been useful in this study. Other examples are the use of population density as a surrogate for potential for farm-gate-sales and population size for market demand at urban centres.

Thresholds and Scoring

The results were affected by the selection of thresholds for each of the factors and for the commercial and small-scale models, and its equivalent in the fish yield model, namely the specification of feeding rate × harvest weight combinations. In each case, the thresholds were based on values or indications found in the literature, or in the instance of the determination of the relative importance of the factors for the models, expert assistance was brought in. However, in some cases it was necessary to use data that were already classified, such as the ranges of population sizes of urban centres that were used to make estimates of the potential demand for farmed fish.

In order to keep the results manageable and for ease of interpretation, four classes of suitability were employed throughout the study, from very suitable to unsuitable. In order to minimize subjectivity in the scoring process, the selection of thresholds was always based on an examination of frequency plots of grid cells for each factor and model.

Nominal Area Estimates and Actual Potential

In order to avoid a serious misinterpretation of our results, it is important to keep in mind that our estimates of areas that have potential for fish farming development have to be thought of as encompassing or containing such potential. Not all of the area that we have identified as having potential can be given over to fish farming. In fact, in practical terms, only a small part of it is likely to be available in some places. For example, in some of the areas, especially those with relatively high population densities, the land already may be in other uses that provide returns greater than those that can be realized from fish farming. Another situation makes the area estimates appear to be greater than they are actually likely to be. This is related to the unavailability of digital maps of the protected and reserved areas of several of the countries, including Argentina, Chile, French Guiana, Guyana, Mexico, Paraguay and Uruguay. Thus, in these countries, some of the area that has been identified as having potential for fish farming is undoubtedly reserved for other uses.

INTERPRETATION OF YIELD POTENTIAL FOR SPECIES

Tambaqui and Pacu

Results obtained for tambaqui and pacu are somewhat unexpected for the warmer waters of Latin America because the former species is generally considered to grow more rapidly in waters in the range 25–35°C (e.g., Saint-Paul and Werder, 1980; Saint-Paul, 1989). Although some experiments (e.g., Miyasaka and Castagnolli, 1992) have shown that pacu may grow better than tambaqui in mean water temperatures ranging from 27–30°C, we suspect that the crops/y results obtained for these species may in part be due to the lack of sufficient grow-out data to enable better growth model parameterization for tambaqui.

Pacu does, however, show better tolerance to colder temperatures than tambaqui (Saint-Paul, 1989) and therefore the higher crop/y output registered from Central Brazil southwards for this species compared to tambaqui is perhaps not surprising.

Improvements

The present study is both more refined and sophisticated than the strategic assessment of aquaculture potential carried out for Africa by Kapetsky (1994). The most important refinement was that new data allowed for a fourfold increase in resolution over that used in Africa (i.e., to 5 are minutes), thereby making the results much more usable in order to assess fish farming potential at the national level. Sophistication was added by making direct estimates of yield potential as the number of crops per year possible for four species by integrating elements of the models from POND©. In Africa, only one species was used and only indirect estimates of potential were made.

Other refinements include:

- A pond water balance model that was that used to estimate water loss from seepage and evaporation on a monthly basis in place of water supply via surface runoff from annual precipitation.
- A wide variety of gridded soil data attributes that allowed for better prediction of the feasibility of pond construction and terrain suitability for ponds compared with only two attributes- texture and slope - that were used in Africa.
- A road infrastructure × urban population size combination that made it possible to estimate urban market potential separate from potential for farm-gate sales, whereas only different thresholds of population density were used as surrogates for market demand in the Africa study.
- Modelling of factors important fo**r** the development and operation of commercial and small-scale fish farming that was an advance over the assumption of equal importance among factors in Africa.

Nevertheless, additional improvements can be made, some of which could be carried out with existing data. For example, stochastic weather data could be used to look at best and worst cases for the operation of ponds and for the production of fish. Such analysis could include inter-annual temperature variations as they affect fish growth and production, and precipitation and evaporation as they affect water availability and pond operation costs. Another improvement would be the overlay of the results for commercial farming in order to establish where the best yields for all four species coincide.

Other improvements would require new data or new approaches. Among the most important are:

1. Land cover at 1 km resolution:
   - o to enable better estimates of land actually available for aquaculture; and
   - o to discern the actual kind and distribution of crop types that, in turn, would allow a better inference of availability of agricultural by-products as inputs.
2. A complete set of polygon data for protected areas, to include all of the countries of Latin America.
3. A set of 5-arc-minute precipitation data covering the Caribbean to allow inclusion of the Caribbean island states in the analyses of fish farming potential.
4. Extending the study to:
   - o include more warm-and temperate-water species. However, in practice, even for those species that are most widely cultured and for which the data are most complete, there are still great gaps in data from operational fish farms. Such data should be included in the growth model. A further extension of this idea would be to improve the growth model to investigate polyculture opportunities;
   - o include more intensive fish culture systems, such as flowing water and re-circulating systems; and
   - o include cold-water species.

However, at a point not too far beyond that to which this study has gone, the utility of a continental-level GIS is lost, and it is much better to proceed to a national-level GIS in order to bring more factors into the analysis, and to take advantage of data at resolutions higher than are appropriate at a continental scale.

Verification

The results of the verification exercise suggested that reasonable confidence can be placed on the predictions of the suitability of grid cell "sites" for commercial fish farming derived from the commercial model. However, it has to be kept in mind that the commercial model is based only on five, very general, factors in weighted combination. As is well known, the weights of the factors can change in relation to the species under culture, the intensity of farming adopted and the location of the fish farming enterprise. The commercial and small-scale models address the relative importance of the factors in a very general way at present and there is room for refinement should there be interest in using different species and different culture systems.

At least three possible reasons for the lack of correspondence between predicted Nile tilapia yields for grid cell sites and the presence of operational farms have been explored. Of these, there is evidence to suggest that low crops/y output may be related to predicted water temperatures being less than observed values. It is also likely that the other two factors (i.e., differences in growth potential among tilapia strains and the lower temperature threshold used for Nile tilapia) may have contributed to the low crops/y output. In practical terms, the verification exercise suggests that the yield model is likely to have predicted a lower growth potential for tilapia at altitudes in excess of about 600 m than is actually the case.

CONCLUSIONS

Factors Affecting Development and Operation of Fish Farms

Potential for inland fish farming is high in continental Latin America. The most important factor - urban market potential - scores high across much of the continent. Therefore, the low level of development is not one of lack of physical access to markets that are within easy reach of suitable fish farming areas.

Latin America appears to have a larger relative area potential for inland fish farming than Africa.

Yield Potential of Species

Overall, the results suggest that, in Latin America, there may be only marginal possibilities for commercial aquaculture operations that target a large harvest size of fish, but use low feeding rates. In general, the results suggest a high potential for the high feeding rate × low harvest weight type of intensive commercial aquaculture operations for all species. It is also evident that growth models which account for the temperature preferences of these species can be used to gauge the suitability of their culture in different regions. Availability of fingerlings, suitable feeds, local markets and export potential - among other factors - will, of course, dictate the type of species that should be chosen for a given region.

Overall Potential

From 38% to 60% of Latin America scores suitable to very suitable for small-scale farming of Nile tilapia and carp, respectively. In the same areas, from 0.9 to 1.7 crops/y for tilapia and from 0.9 to 1.8 crops/y for carp can be realized by harvesting at a modest weight.

For the same two species, from 19% to 44% of Latin America rates from suitable to very suitable for commercial farming, and from 1.2 to 2.4 crops/y for Nile tilapia and from 1.2 to 2.3 crops/y for carp can be realized on the same areas by feeding at 75% satiation and harvesting at a moderate weight. Tambaqui and pacu occupy an intermediate position in terms of the surface area that is suitable or very suitable for commercial farming. From 0.7 to 1.4 crops/y for tambaqui and from 1.0 to 2 crops/y for pacu can be achieved from areas that are suitable or very suitable by feeding at 75% satiation and harvesting at a moderate weight.

Verification

Although the verification exercise was carried out only in one country and involved information on farms culturing predominately tilapia, useful insight was gained into the utility of the commercial model and yield model to predict fish farming potential. The results suggest that the commercial model gives quite reasonable predictions for the weighted combination of the five factors that it incorporates. The yield model to a large extent is dependent on good predictions of water temperature. The lack of a complete gridded data set for weather variables may have been a primary reason for predicted temperatures being lower than values reported for actual farm locations. Lower predictions for yield potential may also be due to different strains of tilapia (e.g., with differing degrees of cold tolerance) being cultured. This problem can potentially be circumvented by re-parameterization of the growth model for cold-tolerant varieties of species such as Nile tilapia. Nevertheless, the most important conclusion here is that the commercial model and yield model appear to work well together to give reasonable indications of fish farming potential. Therefore, the main task of improving the predictions is one of refinement, not reconstruction.

Future Applications

Based on the experience gained in this study a number of future applications can be foreseen. Some are considered below.

Potential of Inland Fisheries

In areas where community water bodies - small reservoirs - are common there are good opportunities to increase the fish supply through various kinds of enhancements. Fundamental questions are:

- What is the basic fishery potential?
- By how much can potential be increased using various kinds of enhancements?
- Which water bodies will have the best productive potential?

The growth model provides a way to look at temperature effects on fish yield potential in both unenhanced and enhanced situations. For example, it should be possible to match candidate species for stocking to a temperature regime that favours growth.

In terms of water temperature prediction, a small reservoir is a large pond. Therefore no great difficulties are foreseen in estimating small-reservoir water temperatures

From the viewpoint of water availability, the water balance approach can be used, modified for reservoir size.

From the viewpoint of productive potential, apart from water temperature, soil characteristics that affect water fertility will be important. Generally, these can be inferred from gridded data sets of soils attributes.

4

Interactions between Fish and Aquatic Macrophytes in Inland Waters

INTRODUCTION

Aquatic macrophytes are among those factors which a fishery manager will try to understand and include in his strategies for optimizing capture fisheries in inland waters. Plant species composition, distribution and percentage cover of aquatic plants may determine the fish species composition, individual fish species production, access to fish stocks by fishermen, fishing gear and sometimes also boat access and transport possibilities for getting the fishery product to the markets. Aquatic macrophytes can also be efficient indicators of water quality, and their presence may enhance water quality due to their ability to absorb excessive loads of nutrients. These properties have been used in wastewater treatment as well as in biomanipulation of water bodies for enhancing fish production.

Aquatic macrophytes are present in virtually all freshwater bodies. In lakes they occupy predominantly the littoral shallows, and their importance in a lake ecosystem generally declines with increasing lake size. The portion of the lake occupied by the littoral zone is inversely related to basin slope and to the degree of shoreline regularity, being smallest in large deep lakes with low shoreline development index. Emergent and floating-leaved aquatic macrophytes seldom grow in waters exceeding a depth of 3 m (Canfield and Hoyer, 1992). In small lakes, where the littoral region with aquatic macrophytes occupies a relatively large portion of the lake, there is a relatively low intensity of competitive interaction among consumers that move from the pelagic zone to use littoral resources (Gasith and Gafny, 1998). The cross-boundary movers are typically adult or large fish that move between deep and shallow water for feeding, for cover, and seasonally, for breeding. The littoral residents, on the other hand, are mostly young, or small-sized fish species.

Aquatic macrophytes can influence the lake ecosystem in a number of ways, which fall into three main categories:

(1) limnological effects related to changes in physical and chemical conditions in the water and sediments;

(2) metabolic effects related to production and processing of organic matter and nutrient cycling; and

(3) effect on biotic interactions and community structure related to the role of macrophytes in providing a structured habitat (Gasith and Hoyer, 1998). The importance of macrophytes will decline with increasing size and depth of the water body. The effects of submersed macrophytes on ecosystem processes have been reviewed by Carpenter and Lodge (1986).

While presently eutrophication is often the cause of a change in the aquatic macrophyte community in a particular water body, there are many instances where there are other reasons for such changes. They may take place spontaneously, or under the impact of intensive grazing by birds, crayfish, grass carp, muskrat, coypu, (see Section 3.) or even cattle. Grazing by cattle on aquatic macrophytes has been mentioned for example in the Netherlands, where it is causing the destruction of reed (Best, 1982), and also in Germany (Sukopp and Markstein, 1981). Another reason for change maybe the appearance of an exotic or adventive plant species. Human impact of non-eutrophicating character, such as mechanical or chemical removal of aquatic macrophytes, possibly targeting the unwanted ones, may also lead to a change in the aquatic macrophyte community.

Freshwater fish use aquatic macrophytes for shelter and refuge, as a food source either direct or indirect, in form of epiphyton (periphyton) and associated invertebrates, and as spawning, nesting and nursery sites. Fish species of densely vegetated margins of some water bodies, e.g.

in the papyrus margins of Lake Victoria, are very tolerant of low dissolved oxygen concentrations and high temperature. Sparse vegetation or isolated macrophyte patches also attract some fish which use them to shelter, feed or spawn. Floating meadows in the Amazon, or wind-blown patches of floating plants, such as water hyacinth or *Pistia*, harbour associated fish fauna.

This publication does not claim a complete coverage of the subject. It does not cover interrelationships between fish and mangroves, fish in salt marshes and tidal flats, use of aquatic macrophytes as food or fish feed additives in aquaculture, fish in ricefields, use of macrophytes for removal of aquaculture effluents and in recirculating fish culture systems, and the importance of dead timber in reservoirs as a substrate for fish food organisms. Several topics are covered only briefly: water-borne disease control (only those larvivorous and molluscivorous fish, which also serve as food, are discussed); biomanipulation (the section dealing with this subject focuses largely on the function of fish versus macrophytes); floodplains (the reader is directed to major publications on this topic). Control of aquatic macrophytes is limited to the use of fish, and to the impact of chemical and mechanical control measures on fish.

AQUATIC MACROPHYTES AS A FISH HABITAT

The term "aquatic macrophytes" includes largely flowering plants (angiosperms), some ferns and mosses (pteridophytes) and macroscopic algae (stoneworts, e.g. *Chara*, and the alga *Cladophora*). De Nie (1987) used the following division for aquatic macrophytes, which, expanded by the major tropical and subtropical plants, is also applied in this publication:

A. Aquatic macrophytes rooting in sediment

   i. Emergent aquatic macrophytes: these include angiosperms *Typha, Phragmites, Scirpus, Carex, Acorus, Butomus, Sagittaria Paspalum, Echinochloa, Vossia* etc.

   ii. Floating-leaved aquatic macrophytes: these include *Nymphaea, Nuphar, Nymphoides, Potamogeton, Polygonum, Hydrilla, Cyperus papyrus* etc.

iii. Submersed macrophytes: these include the algae *Chara* and *Nitella*, moss *Fontinalis*, and angiosperms *Myriophyllum*, *Elodea*, *Potamogeton* etc.

B. Freely floating macrophytes

These include: ferns *Azolla* and *Salvinia*, angiosperms *Lemna*, *Eichhornia*, *Pistia*, *Ceratophyllum*, *Hydrocharis*,, etc.

Cowx and Welcomme (1998) listed a number of characteristics of aquatic plants which make them important to fish:

- water purification - a. direct: for example, by oxygenation and conversion of toxic ammonia to usable nitrates; b. indirect: provision of huge surface area for microbes to do the same tasks
- nutrient recycling - removal during growth season, return during senescence
- physical link between water and air for many invertebrates, e.g. larvae and nymphs of caddis flies, mayflies, chironomids, which are food for fish and have aquatic larval stages and aerial adults
- refugia for zooplankton which graze phytoplankton and keep water clear
- cover for huge range of invertebrates, many of which are food for fish
- cover for fish - value and type varies with age/species of fish as well as type of vegetation
- spawning areas and sites of oviposition for many cyprinid and percid fish, e.g. tench, roach, rudd, perch, and also for pike (all fish of the temperate waters of the western Eurasia)
- food source (living and dying) - direct source of vegetarian fish, e.g. grass carp; indirect source via other animals
- affect flow patterns - accrete sediments and deflect flow thus providing quieter waters and faster shallows
- create discrete habitat which is as functional as physical structure.

Aquatic macrophytes are essential if a healthy fish stocks are to be maintained in natural water body. Opuszynski and Shireman (1995) pointed out that some plants or combination of plant species seem to be better fish habitats than others. Plant density is of importance, and the trophic state of the water body also plays an important role in determining what species and what biomass of fish will be present. The more research results become available, the more complicated the aquatic macrophyte-fish interrelationships appear to be.

The shallow marginal littoral with aquatic macrophytes contains a greater diversity of aquatic animals than the offshore open water areas. For example, in Lake George in Uganda, the vegetated littoral contained 19 fish species compared to around 10 species in the mid-lake (Burgis *et al.*, 1973).

Submersed macrophytes have major effects on productivity and biogeochemical cycles in fresh water because they occupy key interfaces in stream and lake ecosystems (Carpenter and Lodge, 1986). Changes in aquatic macrophytes can have major consequences for commercially important fish species. Most macrophytes are rooted, and constitute a living link between sediments and the overlying water. Carpenter and Lodge (op.cit.) found by 1985 seventy-two papers on the effects of environmental factors on macrophytes, but only 22 papers which emphasized the effects of macrophytes on the environment. Since their publication there have been a number of detailed field studies, and an increasing number of field experimental studies on the function of macrophytes in aquatic environment and their impact on the environment.

In the northern temperate lakes fish community structure and stability have been associated with the presence, abundance, species composition, growth form, and structural heterogeneity of macrophytes in various studies (Weaver *et al.*, 1996). The mechanisms that maintain or alter community structure, and their relative importance, differ among communities and vary with spatial and temporal scale. Aquatic and marginal plants may be a key factor determining whether or not there is any fishery interest in the highly modified urban watercourses which are most accessible to the majority of people.

Aquatic macrophytes have great effect on physical environment, especially light penetration through the water column, water temperature, water flow and substrate. Within stands of aquatic vegetation the light intensity quickly decreases with depth, although great differences exist in the degree of light attenuation betwen species.

The vertical temperature gradient within macrophyte stands is as steep as 10°C m^{-1}, compared to gradients of less than 0.2°C m^{-1} in neighbouring unvegetated areas (Dale and Gillespie, 1977).

Diel oxygen changes as large as 8 mg L^{-1} occur in the waters of dense submersed macrophyte stands (Ondok *et al.*, 1984). Oxygen may be released into water by some floating plants, such as water hyacinth from rootlets of which oxygen actively diffuses into water. Submersed macrophytes oxygenate the water more effectively than floating-leaved macrophytes, as solid cover of the latter prevents effective reoxygenation from the air. Decaying macrophytes consume large amounts of oxygen and macrophyte leachate stimulates oxygen consumption by suspended bacteria. Large mats of floating aquatic plants, such as *Salvinia molesta*, *Eichhornia crassipes* and *Cyperus papyrus* will produce almost totally deoxygenated conditions underneath these plants, because they reduce the light penetration and prevent oxygenation of water blocking the water-air interface.

Macrophytes have positive effect on water clarity through a number of mechanisms. These include provision of habitat for plant-associated macroinvertebrates and a refuge for cladocerans, both of which graze on phytoplankton and epiphytes. While food and habitat provided by plant beds support diverse fish populations, piscivores among them reduce numbers of small planktivorous fish, thereby reducing predation pressure on zooplankton. Macrophytes may also suppress phytoplankton growth through the release of alleopathic substances, and through luxury nutrient uptake and enhanced denitrification. Phosphorus may be reduced by certain plant species that oxidize the sediment. In addition, reduced water movement in plant beds can lead to increased sedimentation rates and reduced turbidity. Where light conditions are deteriorating, mechanisms such as the shedding of epiphyte-burdened leaves, and a shift from low-growing species to taller ones may stabilize the plant state (Moss, 1990).

It is now recognized, that through various pathways, submersed macrophytes have a strong positive effect on the resource base of epiphytic and benthic macroinvertebrates. While many macroinvertebrates adjust their behaviour in response to fish by seeking refuge and/or decreasing risky actrivities, the feeding and individual growth rates of many fish species are positively related to the abundance of macroinvertebrates (Diehl and Kornijow, 1998). However, the negative effects of vegetation on the encounter rates of fish and the positive effects of vegetation on the standing stock and production of macroinvertebrates make it impossible to derive a general relationship between the growth rates of fish and the density of submersed vegetation. It is becoming obvious that the vegetation-macroinvertebrate-fish interaction cannot be understood in isolation from other components of the lake food web and from processes that link the littoral zone to benthic and pelagic habitats (Diehl and Kornijow, 1998).

Extensive macrophyte growth in lakes and reservoirs can alter trophic interactions. Typically, there is a negative relationship between abundance of submersed plants and planktonic algal biomass (Canfield *et al.*, 1983). Factors causing this include: competitive uptake of nutrients by macrophytes; development of extensive algal epiphytes on macrophytes; and a decline in nutrient recycling as macrophytes reduce mixing and nutrient resuspension. Productivity of the epiphyte complex may reach up to 93% of host macrophyte productivity (Jones, 1984). Fish that feed directly on planktonic algae could be negatively affected by dense macrophyte coverage. Aquatic plants can also affect fish populations by altering the abundance and composition of zooplankton and macroinvertebrates, by directly influencing predator-prey interactions, and by increasing the reproductive success of some predators (Maceina *et al.*, 1991).

While aquatic macrophytes represent an important habitat for fish, herbivory by fish, birds, invertebrates and mammals is important in determining macrophyte abundance, diversity and productivity. Herbivory may influence the timing and magnitude of nutrient turnover by macrophytes, water quality and other aquatic biota, including fish.

Aquatic Macrophytes as Preferential Habitat of Fish

Nearly all the fish species connected with littoral zone of European lakes consume some aquatic plant material during their lives. Prejs (1984) listed 15 species, of which ide *(Leuciscus idus)*, roach (*Rutilus rutilus*), and rudd (*Scardinius erythrophthalmus*) had a high content of plants in their food; bream (*Abramis brama*), white bream (*Blicca bjerkna*), chub

(*Leuciscus cephalus*), crucian carp (*Carassius carassius*), gudgeon (*Gobio gobio*), stickleback (*Gasterosteus aculeatus*), tench (*Tinca tinca*), and carp (*Cyprinus carpio*) had small amounts, and bitterling (*Rhodeus sericeus amarus*), bleak (*Alburnus alburnus*), perch (*Perca fluviatilis*), and juvenile pike (*Esox lucius*) had sporadic occurrence. The main plant consumers, roach and rudd, use the majority of the species of submersed macrophytes which occur in sufficiently high densities, but they prefer soft submersed plants. The absence of some plant species in the food, even common species, is due primarily to their physical properties or to their odd taste or toxic properties, e.g. Ranunculaceae.

Several indigenous Euroasian fish species inhabit aquatic macrophytes as adults: tench, rudd and pike. The submersed macrophytes form important feeding habitats for perch, roach, bleak, and eel (*Anguilla anguilla*), while the ruff (*Gymnocephalus cernua*) prefers the *Chara* beds (de Nie, 1987).

In the southern USA, the fish *Enneacanthus gloriosus, Lucania goodei, Lepomis macrochirus, Heterandria formosa,* among others, are substantially more abundant in vegetation than other fish species (Hoover *et al.*, 1989). In the South African lake Swartvlei the fish species *Monodactylus falciformis* and *Rhabdosargus holubi* directly depend on periphyton and associated macroinvertebrates of submersed aquatic plants (Whitfield, 1984). The Australian tropical lungfish *Neoceratodus forsteri* feeds on *Hydrilla verticillata, Vallisneria spiralis* and some other aquatic plants (Kemp *et al.*, 1981).

Among the important freshwater culturable fish species of India, Jhingran (1977) named *Barbus carnaticus* and *Etroplus suratensis* as partially feeding on aquatic macrophytes.

In plant beds rich food supply in form of periphyton, its associated invertebrates and bottom fauna, is generally more numerous and more diverse than in open waters. Within the diverse aquatic plants, the habitat-specific differences in invertebrate composition directly influence habitat selection by fish and foraging behaviour of individual fish (Werner, *et al.*, 1981).

Many young fish need aquatic macrophytes as shelter and protection from predation or to avoid cannibalism. Aquatic macrophytes also serve some fish as a spawning habitat, for the attachment of eggs, and some fish form nesting sites among the macrophytes.

Recent habitat restoration programmes conducted in the Great Lakes Basin of northern America have focused on the importance of aquatic macrophytes for the production of northern pike. It was found that macrophytes are of a greater significance as nursery rather than spawning environment. Adult fish abundance was found to be related to the extent of macrophyte cover, which was found optimal from 35 to 80% but inversely related to body size (Casselman and Lewis, 1996), this resulting in a lower fish biomass in the stands of heavy density.

Significance of Density and Structural Complexity of Aquatic Macrophytes for Fish

A thorough study by Killgore *et al.* (1991) addressed the habitat value of aquatic plants for fishes in North American water bodies, but more specifically in the Potomac River. Their general conclusions can be widely accepted for numerous water bodies in temperate regions of the northern hemisphere. They followed the seasonal change in density and species composition of aquatic plants in relation to the spatial and temporal distribution of fishes. In their discussion, they included references to findings of other authors, supporting their conclusions.

During the spring when plant density is relatively low in the Potomac River, fish abundance is highest in areas without plants because of the occurrence of anadromous, pelagic-oriented species, or in areas with mature stands of *Myriophyllum*, rather than emerging short stands of *Hydrilla.* During the spring, fish tend to associate with aquatic plants that overwinter or emerge early from the substrate for food and cover and disperse to the more established macrophyte beds in the summer and fall.

Once the plant community reached its maximum density in the summer, plants capable of establishing dense stands, such as hydrilla, can occupy the entire water column. This slows down fish movement and foraging efficiency of fish. Therefore, more fish are often found in areas of intermediate levels of structural complexity, particularly for piscivorous species. In late summer through autumn immediate densities of submersed aquatic plants usually contained more species and numbers of fish than areas having dense growths. Some fish species such as pumpkinseed fish (*Lepomis gibbosus*), however, prefer high plant densities. The piscivorous largemouth bass (*Micropterus salmoides*) and yellow perch (*Perca flavescens*) were commonly observed occupying small spaces devoid of plants in dense vegetation, ready to ambush their prey. Banded killifish were found at the water surface immediately above the plants, at the periphery of the dense stands, in spaces ("holes") formed in the plant beds, or the bottom directly below dense canopy formations to utilise both open and structurally complex areas for foraging and predator avoidance. The presence of largely small fish in dense stands of aquatic plants, and larger, piscivorous fish being more common at the periphery of the plant beds, as reported in the literature, was not confirmed from the Potomac River.

Cailteux *et al.* (1994), who compared the growth of largemouth bass in vegetated versus non-vegetated lakes in central Florida observed that bass from lakes without aquatic macrophytes became piscivorous at the size of 60 mm which resulted in significantly faster first year growth than bass from vegetated lakes which did not become piscivorous until when larger than 120 mm. Although largemouth bass from unvegetated lakes grew faster than in vegetated lakes, survival to recruitment appeared to be much lower.

Studies of European fish species have shown that differences in stem densities of aquatic macrophytes result in different foraging rate: the foraging rate of rudd (*Scardinius erythrophthalmus*) on *Daphnia* decreased at stem densities greater than 200 m^{-2} and that of juvenile perch (*Perca fluviatilis*) did not decrease even at the highest stem density of 600 stems m^{-2} (Winfield, 1986). The foraging efficiency of roach (*Rutilus rutilus*) decreased substantially even at the lowest stem density. Similar results were obtained by Diehl (1988) in a study of the foraging efficiencies of roach, bream (*Abramis brama*), and perch feeding on chironomids: the foraging performances of roach and bream decreased strongly in the presence of vegetation whereas the foraging performance of perch was only slightly affected by vegetation. Because perch and roach make up most of the total fish biomass in many European lakes, the effects of vegetation structure on the interactions between these two species will have ramifications for overall community and lake ecosystem dynamics. Roach as a competitor with juvenile perch may severely limit the recruitment of perch to large piscivorous stages in the absence of vegetated habitats (Persson and Crowder, 1998).

In other experiments, when simultaneously offered perch and rudd in field enclosures, pike captured more rudd than perch in environments lacking vegetation, whereas the opposite was the case in environments with vegetation (Eklov and Hamrin, 1989). Pike has been found to be a more efficient predator than perch and pikeperch (*Stizostedion lucioperca*) in vegetation.

Crowder and Cooper (1982) who, in experimental ponds in the state of Michigan, USA, investigated the relationship between macrophyte density (dominated by *Ceratophyllum demersum*) and the interaction between bluegills (*Lepomis macrochirus*) and their invertebrate prey noted that at high macrophyte densities fish captured less prey and grew more slowly despite the presence of a higher biomass of invertebrates. Fish diets at high macrophyte density did not broaden; the fish simply consumed less (but larger prey, than at low macrophyte densities) and grew more poorly as a result. At low macrophyte density fish diets tended to be narrower. The authors concluded that in the long term, intermediate macrophyte density may be the best habitat both in terms of foraging and growth of the predator and stability of the fish-prey interaction. However, they also stressed the importance of considering multiple variables (food, habitat structure, temperature, predators) and assessing resource availability, if we are to better understand resource use in spatially complex environments.

Randall *et al.* (1996) who studied the density of fish in relation to the density of macrophytes in the Great Lakes of North America, found fish production significantly higher in littoral habitats with abundant submersed macrophytes, as measured by percent bottom cover, than in adjacent areas with low macrophyte abundance. Vegetated sites had higher densities of fish, smaller fish, and greater species richness than unvegetated sites. The authors recommended that conservation should be directed at protecting not only the productive vegetated littoral habitat but also adjacent areas which benefit from the high productivity as well.

The architecture or spatial arrangement of plant species has an important impact on fish frequenting aquatic macrophyte stands. Structural characteristics unique to aquatic plants create spatial complexity in aquatic habitats which are important to growth and survival of freshwater fishes. The littoral of lakes and ponds with aquatic macrophytes is physically complex and spatially patchy. Dibble and Killgore (1994), using literature information, listed the following salient points showing the significance of plant structural complexity, i.e. habitat at the microlevel, for freshwater fish:

(i) it deters predation by altering the outcome of predator-prey interactions;

(ii) serves as critical refuge sites for smaller fishes;

(iii) is important for successful spawning of fish;

(iv) increases survival and recruitment of juvenile fishes;

(v) growth rates of young fish increase in these habitats because plant leaves and stems offer a habitat rich in food resources for fish by providing attachment substrate and protection for many microinvertebrates.

Caffrey (1993) summarized the significance of aquatic plant complexity as follows: broad and dissected leaves provide abundant cover and concealment for adult fish and nursery habitats for fry. They also provide a wide range of microhabitats for fish food invertebrates and periphyton, and spawning substrates for fish and invertebrates. Coarse fish which deposit their adhesive eggs on most plant species prefer submersed plants with broader or complex forms.

Floating-leaved plants offer minimal cover, direct or indirect food supply or spawning substrates for fish or invertebrates. This is reflected in the low mean fish standing crop. Filamentous algae do not provide cover, but harbour a large number of invertebrates. This probably explains the good fish crops in Irish canals with a moderate algal growth.

Plants with a year-round presence, such as submersed *Hippuris vulgaris, Myriophyllum spicatum* and charophytes, all broad or complex-leaved, provide favourable conditions for fish and have a high standing crop. However, plants which in a temperate climate die down for the winter months and initiate their growth cycle late in the spring, provide poor habitat for fish (Caffrey, 1993).

Habitat that is spatially complex may incorporate a variety of microhabitats and thereby support a more diverse community. But special complexity may also restrict access to preferred habitat and thereby limit species distribution. Thus, patchy distribution of habitat is important in structuring and stabilizing communities. Even macrophyte beds that are evenly distributed around a lake often are locally patchy, consisting of patches or bands of vegetation that are separated by open areas. Weavei *et al.* (1997) examined how the patchy distribution of submersed macrophytes affects the distribution of littoral fishes within Lake Mendota, Wisconsin (USA). When patchiness in the structure of macrophyte beds was considered along with macrophyte abundance, the distribution response of fish to habitat complexity was better explained.

Patchiness within vegetation provides prey fish with a habitat in which shelter is in close proximity to open spaces that harbour zooplankton, the preferred prey of many age 0+ fishes. Reduced density within the macrophyte patches increases the manoeuverability of prey fish, and the visibility of macroinvertebrates, and thereby improves their capture success on their prey (Dionne and Folt, 1991). In Lake Mendota decline in the abundance of native plants and small fish and in the growth rate of prey and piscivorous fish coincided with the invasion of *Myriophyllum spicatum* in the mid-1960s. During the decline of *Myriophyllum* in the late 1970s native plants have recolonized, and with them the fish community has been recovering (Weaver *et al.*, 1997).

The plant structure complexity also determines the invertebrate biomass in plant stands. In Lake Krankesjon in Sweden *Chara tomentosa*, which grows there in dense beds and constitutes a more complex structure than *Potamogeton pectinatus*, another common plant in this lake, has a higher invertebrate biomass (Hargeby *et al.*, 1994). The complex structure also offers a more efficient shelter against predation from fish (Diehl, 1988). *Chara* overwinters with green parts and thus offers a more permanent habitat beneficial for the aquatic invertebrates, and consequently also more beneficial for fish than *P. pectinatus*, which enters senescence during the long winter period (Hargeby *et al.*, 1994).

The predation by fish on in-benthos (i.e. tubificids and chironomid larvae) is also affected by the density of vegetation. In experiments in enclosures with various densities of floating-leaved macrophytes (nymphaeids) the abundance of chironomids in the upper sediment layer was negatively affected by perch at low and medium *Nuphar* densities (Kornijow and Moss, 1998). Densities of Tubificidae were similar in the surficial sediments in the presence and absence of perch, but in the deeper layers were about twice as high in the presence of fish than in the controls. At high vegetation density perch had no impact on chironomid larvae in the benthos. The authors concluded that the behavioural responses of chironomid larvae and tubificid worms to perch differ, and that a high density of floating-leaved macrophytes protects the chironomid prey from fish.

In addition to the structural benefits provided by aquatic macrophytes for fish, the shade that is created by plants also attracts and may benefit fish. Availability of aquatic plant species that provide appropriate shade may increase both foraging efficiency and predator avoidance by offering a microhabitat that improves detection of food items and predators, which ultimately increases growth and survival (Dibble and Killgore, 1994).

Substrate type and submersed vegetation are important factors in cichlid habitat preferences (Lowe-McConnell, 1987). The American cichlid species occupy a great variety of habitats but prefer muddy bottom areas with abundant aquatic vegetation (Resendez-Medina, 1981). In Lake Bacalar in the Yucatan Peninsula, Mexico, cichlids were found exactly in such a habitat, probably because of the abundance of prey and shelter (Gamboa-Perez and Schmitter-Soto, 1999).

Feeding Habitat

While herbivores may directly feed on aquatic macrophytes, to other fish aquatic vegetation provides a high density invertebrate prey and periphyton (epiphytic algae) associated with the macrophytes. Fish have feedback effects on vegetation by their direct consumption of vegetation and indirectly via other trophic components or abiotic routes such as sediment-feeding induced turbidity. Fish may therefore also affect habitat structure (Persson and Crowder, 1998).

Artificial, macrophyte-like structures were found to impair the foraging of juvenile roach), but were no obstacle to feeding on cladocerans by young rudd and perch (Winfield, 1986). In a study in a hypereutrophic pond in England, Venugopal and Winfield (1993) found that juvenile cyprinids (roach, rudd) and perch were found both among dense, peripheral macrophytes, represented largely by water lilies (*Nuphar lutea*), and in the open water of the pond, although the cyprinids were apparently more abundant among the macrophytes. The higher abundance of cyprinids was confirmed by a significantly higher CPUE within the macrophytes, indicating that they were more abundant or more active in this highly structured habitat. Winfield (1986) found that a dense stand of artificial *Nuphar* sp. had no effect on the feeding rate of zooplanktivorous roach and actually increased the performance of rudd and perch relative to those in open water, possibly due to the reduced predation.

Perch feed effectively on cladocerans in weed beds (Winfield, 1986), and the fish also find a refuge among the plants. Some submersed aquatic plants, such as *Potamogeton berchtoldii*, provide some refuge for zooplankton against predation but this is not the case with the nymphaeid *Numphar lutea* (Timms and Moss, 1984). Further studies of Moss *et al.* (1998) have shown that predation of perch on *Daphnia* increases with less dense stands of the water lilies. Numbers of other small crustaceans and rotifers were mostly unaffected by fish predation. Moss *et al.* (1998) suggested that the refuge will operate even more effectively with cyprinids such as roach and bream who are not as effective hunters among plants structures as perch. Perch may also find alternative food among the plants, such as benthic/epiphytic invertebrates. If these animals are easier to catch than mobile ones in dense vegetation, then this may give an additional refuge mechanism for forms such as *Daphnia*.

A prolonged aquatic macrophyte senescence may have a negative impact through the removal of shelter and food base for fish. It can be compared to a management measure using mechanical control of aquatic vegetation. In the warm temperate coastal lake Swartvlei, South Africa, a littoral plant community dominated by *Potamogeton pectinatus* and two charophyte species underwent a marked seasonal regression in winter months. This resulted in a 60% decline in primary production,

74% slump in littoral invertebrate biomass, and a 54% decline in the abundance of two fish species associated with macrophytes, for which invertebrates and filamentous algae from the littoral zone were of major importance as food source. The disappearance of the macrophytes resulted in a decrease in the condition of *Monodactylus falciformis* and *Rhabdosargus holubi* (Whitfield, 1984). The virtual disappearance of aquatic macrophytes also had an impact on the physico-chemical features of the aquatic environment.

Spawning and Nesting in Macrophytes

It is not the purpose of this publication to give a detailed account of all freshwater fish species which use aquatic macrophytes for attaching their eggs or making nests. Therefore, only some examples are given here. Obligatory and non-obligatory plant spawners (phytophils) stick adhesive egg envelopes to submersed live or dead plants. Among these are many cyprinids such as *Labeo* and small *Barbus*, characin and siluroid fish species, *Puntius (=Barbodes)* spp, *Polypterus* spp, *Clarias batrachus* (Welcomme, 1985). *Gymnarchus niloticus* make floating flask or raft shape masses of aquatic weeds in which they lay their eggs. The siluroids *Callichthys callichthys* and *Hoplosternum littorale* both make a raft constructed of bubbles and aquatic plants. Many cyprinids lay their adhesive eggs on floodplains on flooded terrestrial vegetation. In Africa, *Heterotis niloticus* builds nests in flooded plants at the margins of floodplains. Catfish *Pangasius sutchi* scatters its eggs among submersed weeds and bushes in the shallow inundated margins of the klongs or canals of Indochina.

The following European fish release their eggs among submersed macrophytes where they adhere to the plants: *Tinca tinca* and *Scardinius erythrophthalmus* spawn for example among *Myriophyllum* beds, carp *Cyprinus carpio* utilises a range of plants, including *Carex, Glyceria*, fresh shoots of *Phragmites* and *Salix* roots. Other species include: *Blicca bjoerkna, Carassius carassius, C. auratus gibelio, Rutilus rubilio* and *R. arcasi*. Some other fish stick their eggs to submersed plants but other substrata are utilized in their absence: *Rutilus rutilus, Leuciscus idus, L. leuciscus, L. souffia, Alburnus* alburnus, *Rutilus frisii, R. lemmingi* (Cowx and Welcomme, 1998). *R. rutilus* spawns on *Fontinalis* and *Elodea* beds, *Scirpus, Salix* roots, stones and submersed logs. *Abramis brama* may spawn on stones in lakes, in rivers it utilizes only areas with a weak current and macrophyte substrata, e.g. *Rorippa, Botumus, Sagittaria, Glyceria* and *Nuphar*. Pike (*Esox lucius*) spawns among aquatic plants, such as *Equisetum, Phragmites, Carex* and littoral Poaceae in Finland (Hakkari and Bagge, 1984).

Many North American fish are obligatory plant spawners: these include members of Amiidae (e.g. *Amia*), Esocidae (e.g. *Esox*), Cyprinidae (e.g. *Cyprinus, Cyprinella, Notemigonus*), Catostomidae (e.g. *Ictiobus*), Cyprinodontidae (e.g. *Fundulus*), Atherinidae (e.g. *Labidesthes*), Umbridae (e.g. *Umbra*), Centrarchidae (e.g. *Elassoma*), and Percidae (e.g. *Perca,* some *Etheosoma*) (Pflieger, 1975; Robinson and Buchanan, 1988). Most data on spawning success relative to structural complexity and plants come from studies of adult sport fish that construct nests (i.e. largemouth bass and bluegill).

Although nest spawners successfully spawn in areas devoid of vegetation, they prefer sites with aquatic plants or some other type of structure nearby for refugia. Sunfish of the USA (Perciformes, e.g. *Micropterus salmoides, M. dolomieu*, and *Lepomis gibbosus*) form nesting sites among the macrophytes. Aquatic vegetation is used as nursery habitat for larvae by at least 12 families of the North American fish (Dibble *et al.*, 1996). Some species exhibit ontogenetic shifts in habitat use. For example, prolarvae of yellow perch prefer shallow, dense macrophyte areas, while postlarvae prefer deep, low density macrophyte zones (Gregory and Powles, 1985).

Many phytophils and phytolithophils lay eggs on substrates just below the water substrate where they are vulnerable to sudden falls in water level. Removal of instream vegetation during the spawning period can have a dramatic effect on spawning success. Mechanical control of aquatic plants in streams can also directly remove the eggs of phytophilous fish species and food organisms for the fish.

In Lake Ontario fish which lay eggs in submersed stands include northern pike (*Esox lucius*), perch (*Perca flavescens*), and largemouth bass (*Micropterus salmoides*) (Crowder and Painter, 1991). In five marshes near Toronto, Stephenson (1989) found over 20 fish species using the wetland during the breeding season.

The nesting of the American Perciformes may lead to changes in the aquatic macrophyte species set up. After clearing *Myriophyllum* for their nests and the nesting period, the nests of *Lepomis gibosus, Micropterus salmoides* and *M. dolomieu* have been gradually invaded by diaspore-propagated plant species, including *Isoetes*, which do not spread vegetatively. Every following spawning season about 20% of the original nest sites are not occupied, and the result is a mix of *Myriophyllum* and the other species of aquatic plants which occupy the abandoned nests. The nesting contributes to the formation of a mosaic of different aquatic plant species (Carpenter and McCreary, 1985).

In South America, *Serrasalmus spilopleura* spawn on floating vegetation including *Eichhornia crassipes*, for example in southeastern Brazil. *Serrasalmus* is an important food fish for riverine people in several places in central and northeastern Brazil (Sazima and Zamprogno, 1985).

FISH AND OTHER VERTEBRATES OF FISHERY SIGNIFICANCE THAT FEED ON AQUATIC MACROPHYTES

This section deals only with fish species in which aquatic macrophytes constitute part of their diet. It does not deal with species which, as adults, feed on phytoplankton or filamentous algae only.

Opuszynski (1992) and Opuszynski and Shireman (1995) classify herbivorous fish as those in which food constitutes more than 50% plant material by weight or volume, at least in some period of their life. Most species are opportunistic, changing their diet to whatever food items are abundant or available. Food habits also change with the age of fish and seasons, which may determine food availability. A generalised conceptual model gives major reasons why a herbivorous fish selects its food. Opuszynski and Shireman (1995) mention that 24 families of fish contain freshwater herbivorous representatives. In total 37 species are listed as feeding on macrophytes, and two on "plants material". Twenty of the macrophyte-feeding species belong to the family Cypriuidae, eight to the family Cichlidae. Among these, grass carp is the only fish used on a large scale for aquatic weed control. The authors do not indicate whether under macrophytes are meant aquatic macrophytes, or terrestrial macrophytes (such as flooded grass and other herbaceous plants of floodplains), or both.

Submersed, emergent and floating-leaved aquatic macrophytes are all subject to substantial grazing losses, which often alter relative abundance of macrophyte species and sometimes affect plant productivity (Lodge, 1991). Prejs (1984) gives examples of intensive grazing on *Elodea canadensis* and *Ceratophyllum demersum* by rudd and tench in Polish lakes. Most fish are opportunistic, feeding on a wide variety of food items according to their availability. This is also true of herbivorous fish which frequently ingest aquatic invertebrates dwelling on aquatic plants. The reason is that many aquatic grazers prefer bacteria and microscopic algae because these groups have higher protein and lower cellulose and lignin and ash contents (Lamberti and Moore, 1984).

This is valid for grass carp which feeds on submersed, floating and emergent plants, as well as, e.g. mahseer (*Tor tor*), which scrapes moss and algae from rocks in rivers of India and Pakistan, or feeds on aquatic macrophytes. Opuszynski (1992) pointed out the scarcity of information on the proportion of animal food in herbivorous fish, mentioning microorganisms as the most protein-rich food in nature, and easier to digest than plant cells. Juveniles of herbivorous fish depend on high-protein animal food to survive, shifting gradually to mixed animal-plant food as they develop and grow. Opuszynski mentioned that when grass carp were fed only lettuce in experiments, they were hardly able to grow, but considerable improvement in growth was achieved when the fish were fed earthworm diet.

Plant food is less digestible than animal food because of the cellulose, which is difficult to break down. Grass carp has a muscular gizzard stomach where plant material is ground together with sand grains. Other herbivorous fish use pharyngeal teeth (pharyngeal mill) situated in the throut, to cut, tear and grind the plant material Both methods assist in breaking down the plant cell walls. Herbivorous fish compensate for the shortage of protein in plants (duckweed contains 31%, water hyacinth 24%) as compared to animal food (up to 80% protein) by ingesting higher quantities of food, the daily intake of which, in grass carp, may exceed their body weight. Many aquatic macrophytes have secondary compounds that may reduce grazing. Lodge (1991) in a brief summary of the literature, listed the major groups of compounds: alkaloids, flavanoids, steroids, saponins, phenolics (including tannins), glucosinolates. The digestibility of protein in plants decreases with increasing concentrations of tannins (Mitchell, 1974). Plants containing more than 6% to 7% tannin (dry weight) would be so low in digestibility as to be of little food value; most higher aquatic plants contain less than 2% to 3% tannin when harvested before maturity (Boyd, 1968), but tannin values tend to increase with age.

This chapter leans especially on the information contained in papers and publications by Schuytema (1977); Opuszynski and Shireman (1995); van der Zweerde (1990); Welcomme (1985); grass carp synopsis by Shireman and Smith (1983); special section on grass carp in the United States, as appeared in the Transactions of the American Society 1978, Vol. 107, No. 1; Proceedings of the Grass Carp Symposium (Anon., 1994). This is supplemented by material gleaned from a number of other papers. The section dealing with grass carp is not exhaustive, and the reader is advised to seek more information in the relevant publications listed above. Twelve other herbivorous fish species/genera are dealt with in more detail.

Grass Carp

Adult grass carp eat exclusively plant material, but when young, grass carp are often considered facultative carnivores. When *Tubifex* worms were fed to fry (Haniffa and Venkatachalam, 1980) and fingerlings (Fischer, 1972) of grass carp a large proportion of energy intake was retained. When fingerlings were fed lettuce they grew very little, due to poor digestion and absorption; and when fed pellets made from pondweed they did not grow and had low absorption efficiencies. In comparison, larger grass carp fed *Potamogeton crispus* also had a large faecal loss but retained 12% of the consumed energy (Wiley and Wike, 1986).

While the use of grass carp for biological control of aquatic plants has been found to be advantageous in many situations, especially where chemical or physical control for whatever reason cannot be applied, Schuytema (1977) pointed out at a number of negative impacts, the possibility of which should be assessed prior to the introduction of grass carp. Among them are: uncertainty of the effect on native fish, possibility that removal of plants may eliminate endemic fish food and cover and waterfowl food; lack of knowledge of proper stocking rates; difficulty of live capture; possibility of natural spawning; possibility of introducing a disease vector, lack of knowledge of local plant preferences; and nutrients released into the water by excretion leading to increased primary productivity.

Whatever the negative aspects of using grass carp for aquatic macrophyte control, given existing problems associated with eutrophication and exotic plant species, large-scale aquatic vegetation control programmes using grass carp are likely to increase in the future (Bettoli *et al.*, 1993). In the Netherlands the introduction of grass carp has been considered a success, regarding its control of aquatic weeds (Van der Zweerde, 1990). The area stocked with grass carp has steadily increased since 1977. In 1987 there were 1092 sites with grass carp and 820 users (a total of 270 tons in 2300 ha). Water Boards are the major users of grass carp, with other users having a larger number of relatively small sites.

Feeding Preferences of Grass Carp

Numerous studies have dealt with the aquatic plant preferences of grass carp. Because of food selectivity, the grass carp occasionally increases some stands of plant species by preferential feeding on one or more favoured plants. This situation may continue until the preferred food supply is exhausted, and then it usually starts to feed on less palatable species. Small-sized grass carp utilise soft vegetation while older fish can utilise harder, more fibrous plant material. Water hyacinth is usually avoided, as is the cattail (*Typha*). In some water bodies grass carp may prefer to feed on filamentous algae such as *Cladophora* and *Spirogyra*. Elsewhere, *Nitella* may be avoided. In total, however, grass carp have been reported to consume over 170 different species of aquatic macrophytes (Redding and Midlen, 1992). They also reported that grass carp consumed more softer and succulent submersed aquatic macrophytes when water temperatures were below 12–15°C. Prabhavathy and Sreenivasan (1976) reported that in India grass carp is known to ignore all aquatic vegetation in the presence of *Hydrilla*. Bhukasvan *et al.* (1981) reported that grass carp preferred submersed macrophytes such as *Najas graminae* and *Hydrilla* and the floating macrophytes (e.g. *Azolla pinnata*).

Vincent and Sibbing (1992), who fed grass carp various aquatic plants, followed how the fish chooses and chews its food. Given that leaf toughness, availability of particular leaves, and spacial availability of each particular species of macrophytes are characteristics that affect the energy costs of ingestion, the amount obtained by bite and the potential nutritive value are factors affecting the benefit derived from ingestion (Prejs, 1984). Vincent and Sibbing (1992) detected a clear selection for some plants because of the type of chewing mechanism of grass carp, i.e. pharyngeal teeth. *Elodea canadensis*, grass *Glyceria fluitans*, and filamentous alga *Zygnema* sp. were readily taken. *Lemna minor*, *Potamogeton lucens* and leaves of *Typha latifolia* were not acceptable. The authors suggested the possible limiting factors: *Potamogeton* and *Elodea* were probably avoided as the fish has difficulty in extracting a single, manageable stem. Grass cannot be swallowed quickly enough, and chewing is limiting with *Typha* as the leaves are too tough. *Lemna* is rejected perhaps because it has an unpleasant taste.

Grass carp was introduced into India in 1959, and since then it has been used in composite culture of carps, particularly in ponds infested with aquatic weeds (Nandeesha et al, 1989). It controls there *Hydrilla, Najas, Ceratophyllum, Wolffia, Lemna, Spirodela, Ottelia, Vallisneria, Potamogeton, Utricularia, Trapa, Myriophyllum, Limnophila, Nechamandra*, and algae *Nitella, Spirogyra, Pithophora.* It has been found ineffective in controlling the floating plants *Eichhornia, Pistia, Salvinia* and *Nymphaea.* However, *Eichhornia, Hydrilla* and *Najas* are used by grass carp for attachment of its eggs. Due to the conversion to excreta of aquatic macrophytes using grass carp in composite fish culture reduces the cost of fish production. In India, grass carp has also been used for control of aquatic macrophytes with limited success in the Chambal irrigation system.

In Egypt, experiments with grass carp were carried out in a small drainage canal in the Province of Giza near Cairo, with favourable results: submersed weeds were effectively controlled. Further experiments and management measures covered several other canals in the country. Grass carp also found a ready use as human food, with poaching of fish from the canals on a large scale at night (Van der Zweerde, 1990). The use of grass carp in one of the Egypt's main irrigation canals improved irrigation conveyance efficiency through decreasing the percentage of submersed aquatic weeds from 35 % to 6 % (Bakry, 1996), although not affecting the floating water hyacinth.

In pond experiments in Alabama, grass carp reduced water hyacinth plant size by eating the roots. When stocked at 16 per 0.04-hectare pond water hyacinth was eliminated within 30 days after the fish were stocked (Kilgen, 1978).

In the USA, *Chara* and *Hydrilla* seem to be the favourite submersed plants of grass carp, in Europe it is the exotic *Elodea canadensis*. For more information on plant preferences as food of grass carp see Schuytema (1977) and Shireman and Smith (1993).

The selective feeding of grass carp may lead eventually to the establishment of an unwanted aquatic plant, such as *Ranunculus* in the Karakum canal in Turkmenistan (Charyev, 1984). Triploid grass carp is a functionally sterile strain of grass carp that have proved to be effective in controlling aquatic plants in southern California. Pine *et al.* (1990) followed the impact of grass carp on aquatic macrophytes in static and flowing water in canals in northern California in winter, spring and summer. *Potamogeton pectinatus, Myriophyllum spicatum* and *Potamogeton nodosus* produced longer shoots in canals with flowing water than with static water. During spring and summer, triploid grass carp showed a distinct variation in the preference of aquatic weed types in flowing and static water. Accessibility and ease of mastication were found to be more important in determining preference than nutritional quality of the plants.

Van der Zweerde (1990) enumerated and briefly commented on the factors which influence successful aquatic weed management with grass carp: water depth, regularity of previous control measures, water temperature, grass carp density, size of the grass carp, predation, selectivity of aquatic plants by the grass carp, surface area of grass carp stocking in a channel network, and proper integration of the use of grass carp in the overall management programme of a water body. A successful control of submersed aquatic macrophytes also requires a certain level of dissolved oxygen concentration, with the lethal point for grass carp being 0.5 mg L^{-1}, a level which may be reached for example in dense growth of *Hydrilla* (Sutton, 1985).

Impact of Aquatic Macrophyte Control by Grass Carp on Environment, on other Fish and Waterfowl

When grass carp (stocking rate 74 fish ha^{-1}) eliminated aquatic vegetation over 3 600 ha in the 8 100 ha Conroe reservoir in Texas, this resulted in an increase in nutrient (phosphorus and nitrogen) concentrations, and a decline in water transparency by 40% as a result of a build up of phytoplankton (Maceina *et al.*, 1992). Once the vegetation was removed, autotrophic production switched from macrophytes to phytoplankton, resulting in a nearly twofold increase in algal biomass as measured by chlorophyll *a*. Zooplankton was relatively unaffected by plant control for a year and a half, which was followed by a decline in cladocerans as a result of an increase in planktivorous fish.

Vegetation removal by grass carp in Lake Conroe system represented the single largest reduction in aquatic vegetation by any means in North America (Bettoli *et al.*, 1993). Prior to the biological control using grass carp, the dominant submersed plants were *Hydrilla verticillata, Myriophyllum spicatum* and *Ceratophyllum demersum*. The most notable declines were observed for several small, phytophillic *Lepomis* spp, for bluegill *Lepomis macrochirus*, and for crappie *Pomoxis* spp. Biomass of largemouth bass *Micropterus salmoides* did not decline but the density of age-1 and older fish did decline. Density of threadfin shad *Dorosoma petenense* increased nearly fivefold.

There was a collapse in the stocks of three cyprinodont species and two other species, i.e. bantam sunfish *Lepomis symmetricus*, and brook silversides *Labidesthes sicculus*, while catches of inland silversides *Menidia beryllina* and threadfin shad increased significantly. The original largemouth bass - crappie - hybrid striped bass (*Morone chrysops* × *M. saxatilis*) fishery was replaced by a channel catfish -white bass - hybrid striped bass - largemouth bass - black crappie fishery after vegetation removal. Hence, the structure of the sport fish community changed. Several changes were likely linked to changes in structural complexity and primary production resulting from vegetation removal. The littoral fish community shifted from a sunfish and shad community to one that also included sizeable numbers of cyprinids, inland silversides, and channel catfish. The total number of species declined over the 7-year study period.

Available prey:predator ratio remained similar after vegetation removal for large predators but increased an order of magnitude for small predators. The responses of some species to complete vegetation removal in Lake Conroe are consistent with known life history requirements.

Numerous studies exist on the feeding impacts of grass carp on other fish species. Grass carp consumes phytoplankton and infusoria during the first 2–4 days of life, changing to zooplankton after 5 days. The fish do not become phytophagous until they are about one month old and 2.5–5 cm in length. Aquatic macrophytes become a regular constituent of the diet when the fingerlings reach 55 mm TL (Watkins *et al.*, 1981). In experiments it was found that 7 to 11 months old grass carp would eat aquatic invertebrates and trout fry in the presence of palatable plants when there was no cover for the prey. Elimination of aquatic vegetation may have disastrous consequences on native aquatic invertebrates, fish and aquatic birds, as the balance between plants and the regular local fauna dependent on them is disturbed.

Kilgen (1978), referring to the work of other authors, mentions that grass carp increased bluegill (*Lepomis macrochirus*) reproduction, reduced largemouth bass (*Micropterus salmoides*) populations, and reduced bluegill production in ponds. A study on more of 100 lakes in Arkansas, where grass carp was introduced as a biological weed control agent, showed that plant removal by grass carp did appear to improve the condition factor of largemouth bass, bluegill and redear sunfish (*Lepomis microlophus*), and Bailey (1978) suggested that the introduction of grass carp will not always improve or harm fish populations. This has also been confirmed in the study of 8 lakes in Florida, which have experienced the long-term removal of all aquatic vegetation with grass carp (10 to 15 years): there was no consistent trend with aquatic macrophyte removal. Some grass carp lakes supported excellent fish populations, and some did not. From this study Canfield and Hoyer (1992) concluded that the long-term loss of aquatic macrophytes from a lake ecosystem will not necessarily lead to a decrease in a lake's total or sport fish population, but there exists a potential for such a decrease.

Reviewing the management of aquatic vegetation with grass carp in Iowa over the period 1973–1993, Mitzner (1994) pointed out that in the 2 000 private waters that have been stocked with grass carp, partial control of vegetation is rarely attained: normally there is an all-or-none response. In lakes with 100% control of vegetation there are only minor changes in sport fish populations. According to Mitzner, sport fishing is not adversely impacted by grass carp introductions and in most cases catch has increased because of more available access to shore by anglers fishing for bluegill and crappie.

The beneficial influence of grass carp on tench in the Lancaster Canal, England, is described by Petridis (1990). Grass carp, through moderate consumption of aquatic plants (dominant species *Elodea nuttallii*) produced better conditions for the exploitation of benthic organisms by the benthophagous tench (*Tinca tinca*), because tench (>35 cm fork length) were able to search for food in the open areas created. This also enabled the fish to make use of zooplankton, mainly the large species *Eurycercus lamellatus* whose restricted phytophilic habitat made it vulnerable to predation.

Epibenthic animals, in particular gastropods and *Asellus aquaticus*, suffered from the predation, which caused severe reduction of them in the site totally depleted of vegetation. In this site, tench utilized a diet in which red chironomids were the dominant prey group together with a large amount of fine detritus. Petridis (1995) suggested that a moderate control of vegetation in temperate climates, using the grass carp, could be a satisfactory management technique for the improvement of aquatic habitats. Reviewing literature, the same author mentions an increase in the growth rate of bream (*Abramis brama*) when stocked with grass carp. Benthivorous species are generally considered to increase their production and growth, benefiting chiefly from the direct utilization and nutrient release of the undigested grass carp faeces. In England, tench can be stocked with grass carp at high densities, because both species occupy different ecological niches, and both species benefit each other through feeding activities.

The changes in the natural habitat of some Arkansas lakes (USA) stocked with grass carp resulted in replacement of *Lepomis microlophus* and *Esox niger* by largemouth bass (*Micropterus salmoides*), bluegill (*Lepomis macrochirus*) and the gizzard shad (*Dorosoma cepedianum*) (Baily, 1975). Grass carp stocking also led to a decrease in crayfish (*Procambarus clarkii*) production in small ponds (Forester and Avault, 1978). The interference seemed to arise from the competition for plant food. The grass carp affects other fish through interference with their reproduction, broadening or narrowing their food base, and decreasing refugia (Shireman and Smith, 1983). In the former Soviet Union grass carp introductions have adversely affected pike, perch, crucian carp (*Carassius carassius*), and roach (*Rutilus rutilus*).

In a USA reservoir, the stocked grass carp through the subsequent elimination of vegetation disrupted the macroinvertebrate food base and this resulted in a 50 % reduction of centrarchid biomass (Shireman and Smith, 1983). In the Austrian lake Neusiedler grass carp was introduced alongside six other alien species to enhance the fisheries. Its presence has resulted in almost the complete disappearance of submersed macrophytes, an important spawning habitat for a number of fish species (Mikschi *et al.*, 1996). While it is not possible to separate the impact of this fish from other impacts, such as water level regulatory measures, which led to a loss of habitats within the reed belt due to drying out, and eutrophication, it is believed that its presence contributed to the disappearance of some of the indigenous fish species from Neusiedler Lake.

In Lake Marion, South Carolina (USA), removal of *Hydrilla verticillata* by triploid grass carp led to an increase in littoral fish, especially Centrarchidae, as hydrilla decreased from a maximum of 4 700 ha (approximately 50% of the surface area) to less than 100 ha. The mean lengths of most littoral species were similar during the study period, during which, in spite of a substantial decline in hydrilla, other forms persisted. This provided an intermediate level of structural complexity, preventing any detectable negative effects on the littoral fish (Killgore *et al.*, 1998).

In reservoirs, aquatic vegetation may conflict with other reservoir uses, such as boating, swimming, and electric power generation. Some reservoirs are important habitats for migrating and wintering waterfowl which is foraging on aquatic vegetation. The use of grass carp to control aquatic vegetation in some reservoirs may reduce habitat quality for waterfowl because plants preferred by grass carp are also important as waterfowl food and as habitat for invertebrate food items (McKnight and Hepp, 1995). In Guntersville Reservoir in Alabama, introduced grass carp reduced the presence of native aquatic macrophytes by 44%, but at the same time the invasive exotic milfoil (*Myriophyllum spicatum*) increased by 15%, as grass carp finds this plant less palatable. Although milfoil is eaten by American coots (*Fulica americana*) and some ducks during autumn and winter, a mixture of native plants and milfoil is more desirable. Native aquatic plants provide better habitat than does milfoil for aquatic invertebrates, which are important dietary components for a variety of waterfowl species (Krapu and Reinecke, 1992). Milfoil also is higher in fibre than are native aquatic species typically eaten by waterfowl in southeastern USA and most ducks and geese are limited in their ability to digest and assimilate fibre. Use of grass carp to control aquatic vegetation in Guntersville Reservoir reduced the quality of waterfowl habitat by reducing the biomass of the native aquatic plants.

In Florida Lake in South Africa, which for a number of years experienced a major invasion of *Potamogeton pectinatus*, *Lagarosiphon muscoides* and *L. major*, triploid carp was introduced to control these aquatic plants (Venter and Schoonbee, 1991). Within a period of one year the mean wet mass standing crop of the weeds in the lake was reduced to one sixth of the initial value. Reduction in the weed cover coincided with changes in the populations of aquatic plant-eating birds, such as coot (*Fulica cristata*), Egyptian goose (*Alopochen aegyptiacus*), and yellowbilled duck (*Anas undulata*). Owing to the reduction in plant presence by grass carp, small fish became more exposed to fish-eating birds such as reed cormorant (*Phalacrocorax africana*) and whitebreasted cormorant (*P. carbo*), which have both increased in numbers at the lake following the clearing of aquatic macrophytes from the surface.

Stocking Rates

Stocking rates of grass carp and the age/size of the stocked fish differ from water body to water body and would be determined by many factors, including the amount and species of aquatic plants present, water temperature, and the objective of the use of grass carp as biological control agent. In Russia the stocking rates could be 79 kg ha^{-1} of one-year fish to clear ponds of vegetation, and 254 kg ha^{-1} to keep canals clear of aquatic plants. In India, effective stocking rates ranged from 654 to 5200 ha^{-1} for fish of various weights. In fish culture ponds in the USA a rate of 247–1235 fingerlings ha^{-1} efficiently cleared aquatic weeds from ponds, and in lakes twenty-five of 20–25 cm fingerlings ha^{-1} were sufficient (Schuytema, 1977).

Stocking rates for grass carp for the control of aquatic macrophytes are usually estimated from the surface area of macrophyte coverage. But this reduces the predictability of the outcome because this approach does not account for plant biomass, which can vary with plant depth and the plant species in question (Killgore and Payne, 1984). Osborne *et al.* (1982) determined the stocking rate based on whole-lake plant biomass: 14–20 grass carp per metric ton (wet weight) of plants would eliminate *Hydrilla verticillata*, and the 20 grass carp per metric ton would do this in less than 4 months.

For the lakes in the Pacific Northwest of the USA, where water temperatures are lower than in Florida, Bonar (1990) estimated grass carp stocking rates of 4.7–15.1 fish per metric ton. Cassani *et al.* (1995) tested in eight small warm-water impoundments (0.8–45.3 ha) of Florida various stocking rates for triploid grass carp to identify a threshold rate at which macrophytes were suppressed but not eliminated. Stocking rates of 4–8.4 grass carp per metric ton of vegetation (wet weight) resulted in gradual reduction to zero macrophyte biomass in 8–17 months. The plants in question were *Najas guadalupensis, Hydrilla verticillata, Ruppia* sp., *Chara* sp. (Fig. 6). All these plants are readily consumed or preferred by grass carp, but slight differences in chemical composition of different populations of the same plant species can potentially affect preference and consequently consumption rates. According to Bonar *et al.* (1993) grass carp consumption was positively correlated with the concentration of calcium and lignin and negatively correlated with the concentrations of iron and cellulose.

In the Cassani *et al.* (1995) experiments shifts in macrophyte species composition took place at sites where macrophytes were suppressed or began to regrow after a period of zero macrophyte biomass. However, the number of species present and composing at least 4% by weight of the total macrophyte biomass did not decline. Osborne and Sassic (1981) who studied the impact of grass carp on hydrilla infestation in a central Florida lake concluded that only small grass carp should be stocked, and that one can expect the fish become ineffective as a means of weed control as they approach a body weight of 14 kg, due to their reduction in the absolute, as well as the relative weight of vegetation consumed with increasing size.

In New Zealand incremental stocking of 60 grass carp (total 218 kg live weight ha^{-1}) in a small 1.9 ha eutrophic sand dune Lake Parkinson led to the apparent eradication of *Egeria densa*, and to a more than 99% reduction in the cover and biomass of other macrophytes. Grass carp numbers were then gradually reduced by trammel netting and seining, with the last fish being removed using rotenone. Five years after the removal of grass carp predominantly native macrophytes re-established themselves in the lake. The emergent *Eleocharis sphaceolata* returned to the margins, and the submersed communities dominated by *Potamogeton ochreatus* and *Nitella hookeri* grew to a maximum depth of 5 m (Tanner *et al.*, 1990).

In most water bodies outside its area of natural distribution spawning of grass carp must be induced artificially. However, a number of countries reported natural spawning where the fish has been transferred. In the Tone River in Japan grass carp has spawned since 1947, in a reservoir in Taiwan the escape of fingerlings from a pond resulted in its establisment and natural reproduction. Natural reproduction has taken place in a river and lake system in Mexico after the fish were released following water hyacinth control experiments. In Turkmenistan, grass carp reproduces in irrigation canals. Some countries (e.g. New Zealand) have regulated the introduction of grass carp, and some states of the USA have banned its use.

In some countries the ban on the use of grass carp has been eased under condition that a sterile diploid or triploid form of the grass carp is used. In New Zealand, only sterile fish are released for weed control (De Zylva, 1996). In the laboratory, triploid grass carp is known to exhibit a slight reduction (10%) in ingestion rate (Wiley *et al.*, 1986).

A number of fish diseases has been introduced with the introduction of grass carp into Europe.

Environmental impact of grass carp on the aquatic ecosystem cannot be disputed: by suppressing some species of macrophytes other species may increase; by grazing off some plants the spawning substrata of important fish may disappear; without a counterbalance of phytoplankton-feeding fish, water quality deteriorates. Aquatic systems function best with a moderate abundance of aquatic macrophytes and introduction of grass carp could assist in reaching such equilibrium.

When using grass carp in a multi-use lake (in North America) Leslie *et al.* (1987) made the following recommendations:

i. only sterile triploid fish should be used and they should be used only in lakes with a legitimate aquatic plant problem;
ii. the relative value of different lake uses should be taken into consideration; grass carp are not good candidates for locations where they negatively impact valuable lake resources like waterfowl habitat;
iii. the kind and quantity of vegetation should be surveyed; grass carp may be used where the target species is abundant and highly palatable to grass carp;
iv. grass carp movement should be restricted to target areas;
v. a low stocking rate of grass carp should be integrated with other plant control methods. The best approach is to first reduce the biomass of target plants with other methods and then use a low stocking rate of grass carp to browse regrowth. If stocking rates are too low, additional fish can be stocked.

A grass carp symposium in Florida, USA (Anon., 1994) met to discuss the effects of stocking grass carp for aquatic plant control in the USA. Even after 25 years since stocking grass carp for the purpose of plant control, some management concerns remain open and require better answers. Grass carp has been used especially to suppress the invasive hydrilla, which has rapidly established itself in many water bodies of the USA. Only in Florida, well over 2000 water bodies have been stocked with grass carp. Haller (1994) gave two possible scenarios on the use of grass carp to suppress hydrilla in a landlocked prime sportfishing lake of 2000 ha area. In the first scenario, by the end of the second year, grass carp, after consuming hydrilla, including the regrowth of the second year, starts feeding on native plants, and in the fifth year there are no aquatic macrophytes left in the lake. The lake remains plant free for 20 years, and revegetation is attempted to improve sport fishing, but is unsuccessful because there is a residual grass carp population still in the lake. In the second scenario, Haller shows a combination of application of herbicide and grass carp stocking, but also shows the presence of adverse conditions such as heavy predation on the stocked fish, low water level promoting regrowth of hydrilla, poor impact of herbicide in repeated treatments in the following years.

With further stocking inputs of grass carp to make control more effective, accompanied by rising water level, the result is as in the scenario one. In both situations one ends with a lake with very little emergent vegetation, and no submersed vegetation for 15–20–25 years. According to Haller, at present, there are no clear-cut guidelines for stocking rates of grass carp which could be applied to all the water bodies, and little is known on the long-term effects of grass carp on the ecosystem of such water bodies. At the same symposium Brakhage (1994) pointed out that at present every system will have to be treated separately with a commitment to monitor the effects of the control programme and modify the approach.

The debate will continue over the appropriate amount of aquatic vegetation needed in a lake system, if any, to maintain healthy fisheries. This should also involve considerations of wildlife as numerous birds, reptiles, amphibians, and mammals rely on wetlands for their survival. The majority of these require aquatic macrophytes, and the boundaries of their desirable habitat are generally delimited by the occurrence of aquatic macrophytes within a system.

Despite the extensive use of grass carp for control of aquatic plants, the impact of the control is still not easy to fully predict. Variations in nutrient loading, macrophyte seasonality, climate among the variety of water bodies, as well as variations in consumption rates and preference for certain plant species complicate vegetation management with grass carp, especially where mixed species communities occur (Cassani, 1995). Use of stocking rates that do not account for these differences are largely responsible for the unpredictable impacts of stocking grass carp. Other factors that contribute to the unpredictability of using grass carp for vegetation control are inconsistent survival rates associated with size at stocking, poor water quality, and variable predation pressure. Resource managers will need to place more value on macrophyte suppression rather than elimination so that the low cost of using grass carp is balanced with potentially long-term impacts (Cassani, 1995).

Other Herbivorous Fish

Tilapia Zillii

When 5 to 8 cm long *Tilapia zillii* shifts from predominantly animal to predominantly herbivorous diet. In Africa, *T. zillii* is known to feed readily on *Hydrilla* (Pieterse, 1981). In Lake Naivasha, Kenya, Siddiqui (1977) reported that macrophytes represent 67.7% of its diet, but a more recent information by Muchiri *et al.* (1995) for the same lake has shown that this species is omnivorous. They also compared the food web of *T. zillii* with that of another common Lake Naivasha tilapia *Oreochromis leucostictus.*

In Volta Lake, Ghana, in *T. zillii* and *T. rendalli* higher plants (predominantly of terrestrial origin from flooded land, such as grass) formed 61.4% of the total food eaten (Petr, 1967). In Egypt, *T. zillii* fed upon both plant and animal material. Barnley (undated), who for the Uganda Medical Department prepared a publication on anti-malarial precautions in the management of fish ponds, cautioned that *T. zillii* can eat only leaves of comparatively soft plants, and as soon as they have eaten all, they will starve and die out unless an alternative source of vegetable food is supplied. In fish ponds, if present in sufficient numbers, it will keep the marginal grass eaten down to an extent at which it and other fish (e.g. *Lebistes*) can catch and eat the larvae of mosquitoes.

Preferential feeding of *T. zillii* was observed by Buddington (1979). He found it to prefer *Najas guadalupensis* to *Lemna, Myriophyllum* and *Potamogeton pectinatus.* Saeed and Ziebell (1986) found it to prefer *Chara,* followed by *Najas marina, Elodea densa* and *Myriophyllum exalbescens. T. zillii* avoided bushy twigs or bulky stems of such plants as *N. marina* and *E. densa* and fed on leaves and soft slender stems which are easy to grasp and separate.

In the 1950s the cichlids *Oreochromis leucostictus, Tilapia zillii* and *Oreochromis niloticus* were introduced in Lake Kyoga, Uganda. Lake Kyoga, situated a short distance downstream of Lake Victoria, receives the Victoria Nile. Both species are captured in traps and gill nets set along the marginal papyrus mats or among aquatic macrophytes, while *O. niloticus* supports a very active fishery among floating islands of papyrus (Twongo, 1995). *T. zillii* occurs mainly under the cover of submersed and floating macrophytes such as *Ceratophyllum, Myriophyllum, Potamogeton, Nymphaea* and *Pistia*, and in sheltered bays, often close to the papyrus fringe. The wide use of seine nets has led to the reduction in macrophyte cover and this has been suspected as being one factor which has contributed to the decline in stocks of *T. zillii*. Another factor could be the competition for nursery grounds with other tilapias, i.e. *O. variabilis* and *O. niloticus*.

T. zillii was introduced in the USA for macrophyte control (Shireman, 1984). This species has established self-sustaining populations where water temperature (>10°C) is suitable for overwintering, such as in Arizona and California, or in thermal springs and heated power plant discharges. The results of the impact of this fish on aquatic macrophytes depended on the initial stocking density, the target macrophyte species, and the presence of native fish. The success of the use of *T. zillii* in controlling aquatic plants in ponds with sport fish was found to depend on high survival of tilapia young. Bickerstaff *et al.* (1984) found in experiments with *T. zillii* and the predator *Lepomis macrochirus* that tilapia survival was highest at 75% artificial plant cover. *L. macrochirus* and *T. zillii*, when they appear in the same water body, will occupy the shallow areas near submersed vegetation and are likely to interact with each other. Extrapolation of the survival data from the experiments indicated that too few *T. zillii*, when stocked with predatory sport fish, would survive to the size where they could adequately control aquatic plants. However, at the same time Bickerstaff *et al.* (1984) noted that tilapia fry never retreated into plant cover for protection; they appeared to depend on tight schooling and parental protection for survival.

In a North Carolina (USA) power plant cooling reservoir *T. zillii* rapidly established a reproducing population after inadvertent introduction in 1984. It eliminated the submersed macrophyte community, including a 57-ha infestation of *Egeria densa*, within one year. Just prior to macrophyte disappearance from the reservoir the tilapia density and standing crop were 1 080 fish ha^{-1} and 16.6 kg ha^{-1} respectively (Crutchfield *et al.*, 1992). By 1985, nearly all macrophytes had disappeared, and after

1985 no submersed macrophytes were found in the reservoir. When macrophytes were scarce or absent, tilapia switched to a diet dominated by detritus. The ability of tilapia to change to alternate food permitted the population to continue expanding in the absence of macrophytes. Changes in water quality were minimal after macrophyte removal with no increased nutrient enrichment. Factors leading to the establishment of the tilapia were an overwintering refuge provided by continuous thermal discharge of water temperature above 10°C from the thermal plant, a paucity of predators largemouth bass (*Micropterus salmoides*) and bluegill (*Lepomis macrochirus*), and the species ability to utilise alternate food sources following macrophyte removal.

Macrophytes density decreased with no apparent effect on native fish in some southern California irrigation canals where *T. zillii* was introduced in 1971 (Hauser, 1975). In the Sonoran desert in California (Legner *et al.*, 1975), *T. zillii* and *O. mosambicus* stocked at a high rate were found not to have an adverse effect on the sport fish largemouth bass. As *T. zillii* also consumes sprouting rice, one should very carefully consider its use in irrigation canals bringing water to such fields.

Tilapia Rendalli

T. rendalli (formerly *T. melanopleura*) has demonstrated its potential in controlling aquatic macrophytes by removing them completely from some impoundments (Junor, 1969). However, in some situations the fish has had less impact, especially where the macrophytes are too dense. Higher aquatic macrophytes such as *Hydrilla, Chara, Sparganium, Potamogeton, Leersia, Lagarosiphon, Carex, Typha, Cyperus papyrus* and *Paspalum* have been found in the stomach of this species (De Bont *et al.*, 1949). According to Caulton (1977) the diet of *T. rendalli* is strongly related to the abundance of aquatic plants. In Lake Kariba (Zambia/Zimbabwe), during the period of invasion of this reservoir by *Salvinia molesta*, rootlets of this plant appeared in 19.6% of the investigated stomachs, representing 6.7% of the total food taken, while vegetable detritus represented 60% (Mitchell, 1976). With the decline in salvinia from the 1970s this plant has been gradually replaced by increasing amounts of rooted macrophytes, especially *Lagarosiphon ilicifolius*, which grows down to a depth of around 10m. This has been accompanied by an increase in abundance and biomass of benthic invertebrates, with a spectacular increase in bivalves (Karenge and Kolding, 1995).

In experiments to find out whether *T. rendalli* is a selective feeder giving preference to some plants as compared to others, Chifamba (1990) fed the fish in tanks the following plants: *Ceratophyllum demersum, Lagarosiphon ilicifolius, Vallisneria aethiopica* and *Najas pectinata.* The fish preferred *V. aethiopica, N. pectinata, L. ilicifolius* and *C. demersum* in that order. In the wild, the fish feeding preference was as follows: *V. aethiopica, L. ilicifolius, N. pectinata, C. demersum.* In both cases, *C. demersum* was the least preferred food. The preference of the fish for *V. aethiopica* is explained as given by its high protein and ash content. However, other factors such as palatability, a lot of structural material and polyphenolic compounds, abundance of the particular plant, are also known to affect selection.

Small fish feed on cladocerans, and there is a shift to filamentous algae in fish larger than 50 mm (Munro, 1967). LeRoux (1956) observed a shift away from chironomid larvae at about 130 mm TL. In Malawi, the indigenous fish *Tilapia rendalli* has been used to control aquatic plants in rainfed ponds. The analysis of stomach contents has shown that juvenile fish of less than 150 mm TL prefer to feed on filamentous algae, followed by submersed macrophytes such as *Myriophyllum* and *Vallisneria*, and on softer emergent vegetation (Brummett, 1995). In Malawi, the use of this fish to control aquatic plants in fish ponds was not found to be a practical proposition, as only fish of an already harvestable size could do the job.

In the USA this species is considered potentially harmful because of its competitiveness and reproductive potential. It tolerates salt water and could spread along the coast of Florida from watershed to watershed. It was also feared that this fish could feed on rice seedlings if stocked in irrigation canals. (Schuytema, 1977, and references therein).

In Papua New Guinea, *T. rendalli* was introduced during 1990–92 into the Sepik-Ramu catchment on the New Guinea island with the purpose of enhancing the fish production on floodplains and in floodplain lakes (FAO, 1993). Many floodplain lakes and associated wetlands are rich in aquatic macrophytes. By 1992 the first fish were already entering the local fish markets.

Oreochromis Mossambicus

Lahser (1967) reviewed the previous literature on its feeding preferences: the fish showed its preference for filamentous algae and ability to digest blue-green algae. In his own experiments Lahser (op.cit.) reported that leaves of aquatic plants are removed by *O. mossambicus* to obtain attached periphyton, and that the increase in turbidity caused reduced macrophyte densities. Close observations of feeding fish indicated that the consumption of many aquatic macrophytes was incidental to the removal of periphyton using the plants as a substrate. Leaves, stems, and roots were scraped or rasped to shreds; the plants were killed and consumed secondarily. But *Lemna* and *Azolla* were consumed in amounts equal to those of filamentous algae. *Cabomba, Myriophyllum, Potamogeton, Vallisneria,* and *Najas* were eaten extensively, *Eichhornia, Brasenia* and *Ludwigia* were killed through destruction of roots and stems but were not consumed in any appreciable amounts. Casual observations in the field showed that *O. mossambicus* would eliminate aquatic macrophytes (such as *Najas*, *Heteranthera, Chara*) and marginal vegetation (*Zizania, Setaria, Paspalum, Echinochloa, Cyperus, Polygonum*) from the bottom and margins of rearing ponds. This elimination is effected through grazing and through increasing the turbidity of the water during nest building.

Oreochromis Aureus

Blue tilapia (*O. aureus*), stocked at 500 or 2500 adults ha^{-1} in small ponds in Oklahoma (USA) successfully controlled submersed aquatic vegetation dominated by *Najas* and *Chara* (Schwartz *et al.*, 1986). The speed and degree of control were proportional to initial stocking density, with effective control observed in low density ponds and high density ponds within 120 and 90 days, respectively. Blue tilapia uprooted and deleafed plants, and there was an increase in turbidity, in water temperature and dissolved oxygen levels. The authors also reported a significantly lower stratification in the experimental ponds, perhaps a result of an increase in the wind force on the water surface without macrophytes, and this resulted in a better mixing. Schuytema (1977) cautioned against the use of *O. aureus*, as this species has spread widely through western and central Florida, where it is competing with the native fish species, and dominates the fauna in many eutrophic Florida lakes. In the state Oklahoma, where Schwartz *et al.* (1986) carried out their experiments, this danger does not exist, as the fish generally could not survive the winter temperatures.

In Israel, *O. aureus* has been found to contribute to cleaning water in reservoirs by feeding on organic sediments (Leventer, 1981).

Feeding experiments have shown that *O. aureus* will lose weight when fed pure plant diet (Okeyo, 1989). This has confirmed similar experiments on *O. mossambicus* and *Tilapia zillii*, and it shows that supplemental protein is required to support the growth of herbivorous fish.

Osphronemus Gourami

The gourami (*Osphronemus gourami*) has been considered useful in controlling some submersed macrophytes in Asian ponds and reservoirs. Edwards (1980) reported this species to feed mainly on plant leaves. It was introduced into irrigation wells in India from Java to control submersed macrophytes. In India giant gourami has also a large appetite for *Pistia stratiotes*, on which mosquitoes transmitting filariasis, breed. Full grown gourami consumes 300 g of *Pistia* day^{-1}, and can clear a 1-ha pond in a month (Anon., 1989).

Trichogaster Pectoralis

This species does not feed directly on aquatic macrophytes, but the macrophytes represent an important link in its life cycle and indirect source of food. Mature fish build nests to spawn in the macrophytes. The fry, which hatches within 24 hours, after the absorption of yolk sac will feed on phytoplankton and zooplankton.

The adult fish feed on periphyton and small invertebrates. In Thailand fish ponds aquatic weeds such as *Eleocharis equisetoides*, *Paspalum conjugatum* and *Hymenachne myurus* are cut and serve as fertilizer the presence of which results in zooplankton bloom in a short time (Boonsom, 1984). The fish is an important food species in southeastern Asia, where in the nature it is common in swamps, canals, lakes or in any shallow water and sluggish current. It is also commonly grown in rice fields.

Tor Spp

Pathani (1980) and Pathani and Joshi (1980) found that in India mahseer (*Tor tor* and *Tor putitora*) feed on and control the growth of submersed plants such as *Ceratophyllum demersum, Myriophyllum* sp., *Hydrilla verticillata* and *Vallisneria spiralis.* In the Narmada River, India Desai (1970) found the food of *Tor tor* dominated by macrophytes, feeding mainly on aquatic macrophytes and filamentous algae, but it also consumed molluscs and insects. Desai (1970) believes that mahseer could be a useful fish for controlling both aquatic macrophytes and with them associated molluscs, intermediate hosts of trematodes causing parasitic infection of fish. In Lake Govindgarh in Madhya Pradesh fingerlings up to 160 mm were found to subsist mainly on macrophytes, while adult fish over 200 m preferred animal food, such as insects, molluscs and fish (Pisolkar and Karamchandani, 1984).

A study on the food and feeding habits of *Tor tor* in Meghalaya, northeastern India, where Dasgupta (1990) collected the fish from the Simsang River, has shown that while the larger fish feed predominantly on algae and macrovegetation, in its juvenile stage they consume more insects. Dasgupta does not provide information whether they feed on submersed terrestrial vegetation during floods, or on aquatic macrophytes. In the more recent publication, Desai (1992) compared the food of *Tor tor* from the Narmada River with that of Govindgarh and Bhimtal lakes and the Bhagirathi River. The first two localities are in the plains of Madhya Pradesh, the last two are situated in the Himalayan uplands.

He concluded that in lakes and reservoirs with clear water and the presence of aquatic macrophytes, the fish would do well and should be stocked. Mahseer will also feed seasonally on aquatic plants in lowland rivers, such as Narmada, where the clarity of water during summer and winter favours abundance of macrophytes and algae. In more turbid waters and during floods mahseer will become an opportunistic feeder taking macroinvertebrates, crustaceans, molluscs, attached algae, plant detritus etc.

Puntius (Barbodes) Spp

The cyprinid *Puntius* (=*Barbodes)* species are generally omnivorous, with a tendency towards feeding on plants. Nandeesha *et al.* (1989)

summarized the information available on the feeding habits of four species of *Puntius* in India. *P. pulchellus*, which reaches 8 kg weight in the Anjanapur reservoir in Karnataka and supports there a good fishery, was found to feed on *Cyperus, Typha, Scirpus, Leersia, Pseudorphis, Hydrilla, Vallisneria, Lemna*, and also on the roots of water hyacinth. According to Devaraj and Manissery (1979) this species shows a great promise in controlling aquatic weeds in ponds. Fingerlings stocked in cisterns with *Lemna* and *Hydrilla* fed on them at a rate of more than 50% of their body weight per day. Nandeesha *et al.* (1989) found *P. dobsoni* and *P. sarana* to feed on *Chara, Vallisneria, Hydrilla*, diatoms and green algae. *P. kolus* prefers planktonic algae and plant matter, but also takes molluscs.

In Sri Lanka, the diets of *Puntius amphibius* and *P. dorsalis* consist of 23 and 27 identifiable plant species, respectively, of higher plant leaves and animal matter (De Silva *et al.*, 1980). In both species the major contribution comes from the plant material. In reservoir Parkrama Samudra in the littoral area with dense cover of *Ceratophyllum, P. filamentosus* was the dominant species in fish catches. This fish cuts the plant into pieces between 2 and 5 mm length, which, however, are poorly assimilated as the digestive system of the fish is unable to attack crude fiber. Hofer and Schiemer (1983) have suggested that the fish probably obtains much of its nutrition from the animal, bacterial and algal periphyton. *Puntius sarana*, another important species of the reservoir fisheries in Sri Lanka, appears there to be rather omnivorous than herbivorous.

In Bangladesh, *Puntius javanicus* was effective in controlling aquatic vegetation under experimental conditions. In Indonesia and a few other Asian countries it serves the dual purpose of fish production and weed control.

P. gonionotus, native to Thailand, Malaysia, Laos, Vietnam and Java (Indonesia) is now widely distributed throughout the Asian region, due to its use in aquaculture and introductions to establish commercial fisheries. It feeds on algae and aquatic macrophytes. It is used extensively for weed control in fish ponds (Jhingran and Pullin, 1985). Scattered references to its habits indicate that the species does not deviate from this essentially vegetable matter diet. *Puntius gonionotus* controlled a dense cover of *Ceratophyllum* in a 284 ha reservoir in East Java in Indonesia within 8 months of stocking (Schuster, 1952). In experiments carried out in Bangladesh on the diet and feeding ecology of the introduced *Puntius gonionotus* by Haroon (1998), macrophytes represented 39.2% of the gut content in small fish, and 15.7% in large fish. Haroon classified this species as macrophytophagous column feeder, depending on aquatic macrophytes with increase in size and development of pharyngeal mill, and with benthic foraging on tiny molluscs as the fish grow larger. This supports the findings by Ukkataweat (1979) who found this species feeding on macrophytes in Thailand at the size >12.5 cm.

To achieve fast results in submersed aquatic macrophyte control, *Puntius gonionotus* needs to be overstocked. Overstocking may also be required to prevent that not all this tasty fish end in the nets of commercial and subsistence fishermen prior to achieving the required results. The same concerns other aquatic weed-feeding fish, such as *Tilapia rendalli* and *T. zillii*.

Colossoma Spp

Colossoma, a South American characid (Serrasalmidae), is an important fish in capture fisheries in most tropical South America. Petrere (1983) lists *C. macropomum* as important fish in catches in Amazonas State in Brazil, and Payne (1987) noted that *C. macropomum* and *C. brachypomum* are of major importance in catches from the Rio Mamore (Trinidad). *C. macropomum* is present in the Orinoco River system in Venezuela, and the Amazon system of Brazil, Colombia and Peru. The fish may reach over 30 kg weight and 1 m length. In the wild, *Colossoma macropomum* and *C. brachypomum* feed on plant seeds and fruits in inundated forests, and with the retreat of water they will feed on zooplankton, fish, insect larvae (Goulding and Carvalho, 1982). *Colossoma* can tolerate low concentrations of dissolved oxygen for short periods of time, and a flap on the lower lip, when extended, allows the fish to skim the surface layer of water for more oxygen when necessary (Ginnelly, 1990).

This adaptation enables it to survive through the dry season in pools which become isolated from the river as flood waters recede. Goulding (1980), who studied the food of this species captured from the Rio Machado (Venezuela) flooded forests, found that rubber tree seeds (*Hevea spruceana*) and palm nuts (*Astrocaryum jauary*) were the dominant food consumed. Goulding notes that these may be selected because they are hard and most other fish species cannot exploit them. Other fruits/seeds are probably competed for by hundreds of other fish species. Large fat reserves are built up during the flood season, as the dry season is a time of poor feeding. In varzea, juveniles of the characid *Colossoma macropomum* have been found in the floating meadows to feed mainly on filamentous algae and wild rice seeds (Goulding and Carvalho, 1982).

Araujo-Lima *et al.* (1998) estimated the contribution of the flooded forest to *Colossoma macropomum* production and the economic value of this contribution to the economy of Manaus, the largest city in Central Amazon. Flooded forest seeds were responsible for more than 41% of the carbon assimilated in over 80% of the examined fish, while carbon from aquatic grasses contributed less than 5%. The total gross revenue from the sale of this fish in Manaus in 1993–1994 was estimated at US$ 13 million, US$ 8.2 million (65%) of which came from *C. macropomum* produced with flooded forest carbon.

C. macropomum has been introduced into aquaculture. It will readily eat farm produce, and hatchery operations for breeding and growing this fish are not complicated (Ginnelly, 1990).

This species has been cultured in Venezuela, Brazil, Colombia, Panama, on some Caribbean islands, as well as in some countries of Asia. More information on culture of *C. macropomum* is available for example in Martines Espinoza (1984) and Ginnelly (1990). Some other species of *Colossoma*, such as *C. mitrei*, an important fish of the Parana-Uruguay river system, have also been tested for the suitability for aquaculture (Merola, 1988). After the construction of hydroelectric dams on the Parana River this species has almost disappeared from the river. A restocking of the reservoirs will require that stocking material is produced in hatcheries.

C. bidens, highly adapted to eating fruits and seeds in flooded forest, feeds during the water level decline on leaves and grass, and hence it has a much higher mean stomach fullness in the dry season than *C. macropomum*. This species has been released in the Sepik-Ramu river system in Papua New Guinea with the purpose to enhance the fish production from the extensive floodplains and to make use of the fruits and seeds of the periodically inundated forests (Coates, 1997).

SCARDINIUS ERYTHROPHTHALMUS (RUDD) AND RUTILUS RUTILUS (ROACH)

The best accounts on food and feeding habits of rudd and roach are presented in papers by Prejs (1984), and Prejs and Jackowska (1978). The diet of rudd, one of the most common littoral fish in eutrophic European lakes, includes 65–90% submersed macrophyte tissue (Prejs, 1984). The contribution of submersed macrophytes to the food of rudd in three Polish lakes investigated by Prejs and Jackowska (1978) increased with the size of fish, attaining over 90% of the total food weight of fish longer than 16 cm. According to Van Donk (1998) only larger rudd (age 1+) are herbivorous.

In Polish lakes, *Elodea canadensis* was found most frequently in the food of rudd and roach, although its biomass in lakes was lower than that of one or two or three of the dominant plant species. In Lake Mikolajskie and in Lake Warniak roach and rudd consumed during the period of intensive feeding 775 kg ha^{-1} and 651 kg ha^{-1} of submersed macrophytes, in the respective lakes (Prejs and Jackowska, 1978). This represented some 50% of the total weight of the macrophytes consumed. Second among the macrophytes consumed was *Ceratophyllum demersum*, and third Characeae and *Potamogeton pectinatus*, respectively. Hansson *et al.* (1987) consider rudd a typical periphyton feeder, which also feeds on invertebrates associated with aquatic macrophytes, and on zooplankton. Rudd only graze during the growing season of the macrophytes at water temperature >16°C (Van Donk, 1998), and, if having the opportunity of feeding on a variety of plants, it may avoid *Ceratophyllum*, which is calcareous in structure (Prejs and Jackowska, 1978).

The macrophyte *Anacharis canadensis* was eliminated in experiments in aquaria by the grazing activity of rudd (Hansson *et al.*, 1987). Artificial, macrophyte-like structures were found to impair the foraging of juvenile roach, but were no obstacle to feeding on cladocerans by young rudd and perch (*Perca fluviatilis*) (Winfield, 1986). Winfield found that a dense stand of artificial *Nuphar* sp. usually increased the performance of rudd and perch relative to those in open water. Prejs (1978) interpolated his experiments on the daily intake of macrophyte food in relation to body size of rudd and roach to natural conditions, and concluded that the daily rations decrease linearly with the increasing size of fish. Fish of a body length 11 cm consumed 15.5 % of the body weight, while fish of 24 cm body length consumed only 7.7%. A similar decrease in food intake with age has been observed for example with grass carp. In Polish lakes in the food of larger rudd and roach *Elodea* dominated.

In Lake Zwemlust in the Netherlands the biomass of submersed macrophytes removed by rudd was less than that removed by coots (Van Donk, 1998). Both fish and bird grazing on macrophytes may affect the internal balance among autotrophic components of a water body, by reducing the biomass of macrophytes, thereby reducing their competition with algae for nutrients. Grazing pressure by rudd is, contrary to grazing by coots, unevenly distributed among macrophyte species. Due to the selective grazing by rudd on its preferred food *Elodea* it could have affected the shift in dominance from *Elodea nuttallii* to *Ceratophyllum demersum* (Van Donk et al., 1994).

Coots, however, pull out whole plants and may influence the macrophyte composition and succession by especially removing plants still present during autumn and winter. After such grazing losses to *Elodea*, next spring other macrophyte species such as *Ceratophyllum*, which forms dormant buds during winter, can occupy the whole available area.

Cyprinus Carpio (common carp)

The habit of feeding on bottom sediments, which uprootes aquatic plants and stirs the sediment, which in turn leads to an increase in water turbidity, makes common carp an unwanted species in some water bodies, especially those which serve as a source of drinking water. At a density of 400 carp ha^{-1} in ponds the common carp activity controlled the submersed aquatic plants in Alabama.

At a density of 448 kg ha^{-1} the carp destroyed submersed vegetation in enclosures placed in a Lake Erie marsh. Common carp is also very numerous in shallow bays of Lake Ontario, where it causes resuspension of sediments and uprooting of aquatic macrophytes (Crowder and Painter, 1991). In a 1215 ha reservoir in Arkansas about 10 tons of carp stocked over 3 years controlled *Ceratophyllum demersum* and *Elodea* (Schuytema, 1977).

In India, common carp is considered to be indirectly useful in controlling submersed rooted vegetation because of its roiling habit (Nandeesha *et al.*, 1989). This leads to an increase in water turbidity which in turn suppresses the growth of submersed aquatic plants. The combination of common carp and grass carp was used to controll *Vallisneria* in irrigation canals of Tungabhadra project in Karnataka, and common carp was also used successfully for controlling *Vallisneria* in perennial ponds.

The presence of common carp and large bream (*Abramis brama*) at densities of 500–700g ha^{-1} in some shallow Dutch lakes is to have a major role in increased turbidity resulting from resuspension of sediments (Meijer *et al.*, 1990). The authors constructed a model for assessing the direct impact of fish on the turbidity. According to the model benthivorous fish having a biomass of 600 kg ha^{-1} are able to reduce the Secchi disc transparency to 0.4 m (in absence of algae and at a depth of 1.0 m). Interestingly, in the Netherlands in experimental ponds with benthivorous bream (*Abramis brama*) and carp (*Cyprinus carpio*) it was the bream which resuspended more sediments than the carp (Breukelaar *et al.*, 1994). Although carp can dig deeper and more efficiently than bream the concentration of suspended solids was 50% lower than in ponds with bream.

A possible explanation for this difference is that carp also feeds on molluscs in contrast to bream and does not have to resuspend the sediments for feeding on these organisms. Unlike bream, the carp had no negative impact on the density of chironomids suggesting that it used different food than bream.

In South Africa and in Australia common carp is known to be one of the few fish species tolerant of high suspended sediment concentrations, such as phosphorus-rich colloidal sediments of some South African rivers (Bruton, 1985). In Australia, where many rivers have a persistent high turbidity due to suspensions of dispersed clay particles, Fletcher *et al.* (1985) found no impact of high carp densities on water transparency. The turbidity values at the Australian sites were generally above 50 NTU and could be considered high. In one instance, with a density of carp of 690 kg ha^{-1} the turbidity was found lower than at a number of other sites. In the USA, Robel (1961) also found no significant differences in turbidity in ponds at carp densities of up to 670 kg ha^{-1}. On sandy substrates, much higher density of fish than observed in the Dutch lakes would be required to increase the turbidity (Robel, 1961). However, Meijer *et al.* (1990) suggested that if macrophytes are to become re-established, the reduction in benthivorous fish, such as carp, is essential.

In Australia, where the exotic common carp has spread rapidly through the Murray-Darling Basin since the early 1960s, the residents along these rivers attributed the decline in aquatic macrophytes since the early 1970s to the expansion of carp.

Fletcher *et al.* (1985) identified a critical carp biomass of 450 kg ha^{-1}, above which plant damage was possible, with those most likely to be affected being *Potamogeton* and *Chara.* These impacts are generally similar to those reported in Canada (McCrimmon, 1968) and USA (Crivelli, 1983). Roberts *et al.* (1995) re-examined the potential for carp to affect aquatic ecosystem structure and processes under Australian conditions, using small ponds.

Out of five plants tested, the carp reduced only the abundance of the submersed *Vallisneria* and *Chara*, to zero in eight days. Crivelli (1983) also mentioned that *Chara* is particularly susceptible to carp foraging. The experiments by Roberts *et al.* (1995) also showed that the sediment surface around *Juncus, Schoenoplectus* and *Myriophyllum* were not worked over by carp.

The presence of these plants seemed to effectively inhibit carp foraging in the substrate, hence reducing the turbidity or suspended sediment concentration, and therefore having a stabilizing role in the aquatic environment.

The biomass of fish communities of reservoirs in the central and southern USA are dominated by omnivorous fish such as gizzard shad (*Dorosoma cepedianum*) and common carp. According to Gallo and Drenner (1995) these omnivorous fish act as "nutrient pumps" when they consume sediment-bound nutrients and subsequently excrete them into the water column. Also clay particles with adsorbed nutrients may be resuspended by the fish feeding activities. These activities almost always lead to an increase in phytoplankton biomass and/or primary productivity. In pond experiments with eleven aquatic macrophytes, the presence of carp reduced the biomass of *Najas guadalupensis*, filamentous algae, and the number of macrophyte species.

According to the authors, the presence of the macrophytes may have "dampened" the effects of carp on the water column. Carp-mediated increases in turbidity, total phosphorus, and chlorophyll a appeared to diminish with increased biomass of *Myriophyllum* suggesting that this plant may regulate the effects of carp on the water column.

In Argentina, a high common carp biomass is associated with high turbidity and high conductivity. Apart from the common carp, other fish found in such environment were pejerrey (*Odontesthes bonariensis*), madrecita (*Astianax eigenmanniorum*) and lisa (*Mugil liza*). In southern Argentina the common carp has spread rapidly through the irrigation systems of the lower valley of the Rio Colorado, and it is now found in 5400 km of irrigation canals and 3700 km of drainage canals. In spite of the canals being dried out for 2–3 winter months for maintenance, the fish re-enter the system every year from the river (Fernandez *et al.*, 1998).

Anecdotal evidence suggests that the originally clear water of canals has become gradually turbid under the impact of the invasion and increasing numbers of the common carp.

The turbidity has led to the disappearance of submersed aquatic macrophytes, formerly infesting the canals. In those sections of canals with lowest conductivity and clearest water a high biomass of the plant *Chara contraria* tended to occur, and these sections also had the lowest carp biomass.

Field experiments on the interaction of the common carp with *Potamogeton pectinatus* and other submersed plants in the irrigation canal system found a significant relationship between the reduced plant growth and an increase in water turbidity resulting from the bottom-feeding habit of the fish. In experiments in aquaria the presence of carp increased turbidity from 2.2 NTU to up to 73.4 NTU, which was accompanied by an up to 58% decrease in submersed plants. In the Argentinian irrigation canals the common carp was also found to feed on submersed plants. Fernandez *et al.* (1998) suggested a possibility of manipulating the size and age structure of the carp population of the irrigation canals for aquatic plant management purposes.

In 19 wetlands in the Great Lakes basin of Canada, waters with turbidity above 20 NTU contained less than five species of submersed plants, while a more diverse community existed in less turbid systems (Lougheed *et al.*, 1998). By extrapolation and with support from the results with enclosures with different densities of common carp and monitoring the impact on aquatic environment, the authors suggested that in one particular water body water turbidity cannot be reduced below 20 NTU unless the common carp is excluded.

In experimental ponds in New South Wales, Australia, high density of carp increased turbidity from approximately 7 NTU to 26 and 73 NTU by day 4, resulting in a complete loss of two (*Chara fibrosa* and *Vallisneria* sp.) out of three submersed aquatic plants (Roberts *et al.*, 1995). Under low impact conditions the uprooting rate of *Vallisneria* was reduced to a third.

That common carp is not selective in its choice of substrate for attachment of eggs, has been well documented from the irrigation canals and storage reservoirs of the Murrumbidgee Irrigation Area in New South Wales, Australia. Carp deposited their eggs on dead and live perennial *Persicaria* sp., grass *Pennisetum clandestinum*, milfoil *Myriophyllum papilosum*, cattail *Typha orientalis*, on filamentous algae, but also on stones and broken off fresh eucalypt branches, and on raffia, which was used as an artificial substrate (Adamek and Adamkova, 1999).

OTHER HERBIVOROUS AQUATIC VERTEBRATES

Activities of some other aquatic or semi-aquatic vertebrates, such as turtles, birds, manatee, nutria, muscrat and cattle, through their grazing activity on aquatic macrophytes, may influence fish species composition, abundance and production.

Turtles

The herbivorous turtles *Kachuga tectum* and *Hardella thurgi* feed on aquatic plants in India and Bangladesh (Chokder, 1967), with the first species found to feed in India on the following: *Lemna, Ceratophyllum,*

Eichhornia, Hydrilla, and *Ipomea.* The species can be easily bred under controlled conditions and there is a good scope for their utilisation in the biological control of aquatic weeds (Nandeesha *et al.*, 1989). In Florida, the turtle *Pseudoemys floridiana* is also herbivorous (Yount and Crossman, 1970). In Delaware waters (USA) Townsend (1979) estimated that fresh aquatic macrophytes formed about 13 % of the diet of the snapping turtle (*Chelydra serpentina*).

Birds

Many ducks and geese, domesticated or not, as well as swans and coots, are consumers of aquatic plants. The smaller species of ducks are rather selective, giving preference to duckweeds and other small plants with soft tissues.

Among aquatic birds, only geese and swans are strict herbivores, and apart from aquatic macrophytes both consume a certain amount of terrestrial plants. Other aquatic birds consume seeds and/or fruits of aquatic plants, thus reducing their reproduction (Van Zon, 1976). McAtee (1939) found that a number of waterfowl species feed on Potamogetonaceae, Cyperaceae and Polygonaceae, Potamogetonaceae being the preferred food source. Grazing affects both the biomass and the growth of submersed macrophytes significantly. In the colonization phase waterfowl may prevent growth of submersed macrophytes (Moss, 1990).

Waterfowl often consume significant portions of annual aquatic macrophyte production: 40% peak standing crop of *Potamogeton pectinatus* in Delta Marsh, Manitoba, Canada (Anderson and Low, 1976); 30% peak standing crop of *P. filiformis* in Loch Leven, Scotland (Jupp and Spence, 1977); and about 50% peak standing crop of an assemblage of submersed macrophytes in Tipper Grund, Denmark (Kiorboe, 1980).

Shifts from high density in aquatic macrophytes to phytoplankton and back to macrophytes may occur naturally or under human influence. Such shifts may also lead to changes in species composition, biomass and migrations of aquatic birds. When a storm destroyed macrophyte beds in Lake Ellesmere in New Zealand, the population of black swans declined from 40 000–80 000 to 4 000, over a period of about 20 years (McKinnon and Mitchell, 1994). In southern Sweden in Lake Krankesjon,

Potamogeton pectinatus and *Chara tomentosa* expanded spatially after more than a decade of phytoplankton blooms and sparse submersed vegetation (Hargeby *et al.*, 1994). During the expansion of submersed plants the number of resting and breeding waterfowl increased.

The increase was significant for the herbivorous coot (*Fulica atra*) and mute swan (*Cygnus olor*), and for omnivorous dabbling ducks. Submersed plants, and especially their green parts, form the main food item of swans and coots, and the increase in abundance of these birds indicates that shortage of suitable food was limiting during the previous state of phytoplankton dominance.

The increased abundance of dabbling duck was related to the increase in both plants and invertebrates. Further changes occurred, such as decrease in the density of planktonic Cladocera, increase in the species diversity of benthic invertebrates and the change in their dominant species. The increase in the mean size of perch (*Perca fluviatilis*) could be the result of a better food supply in form of invertebrates associated with the aquatic plants.

Invasion of Currituck Sound, North Carolina, by *Myriophyllum spicatum* increased the total biomass of submersed macrophytes, and this in turn led to an increase in numbers of dabbling ducks (*Anas rubripes, A. platyrhynchos, A. acuta, A. americana, A. crecca*), ring-necked ducks (*Athya collaris*), and American coot (*Fulica americana*) (Wicker and Endres, 1995). The above examples suggest that population densities of water birds are often closely related to the abundance of aquatic macrophytes. But it is also true that an increase in macrophytes that results in more food for some water bird species may represent a decrease in available habitat or feeding opportunity for other species. A negative relationship between macrophyte cover and bird abundance has been recorded for several species of birds associated with Florida lakes (Hoyer and Canfield, 1994).

Nelson and Kadlec (1984), in their review paper describe the interactions occurring among macrophytes, macroinvertebrates, and waterfowl in freshwater wetlands as a complex interdependency in which dynamic changes in the abundance and distributional pattern of macrophytes resulting from climatological fluctuations influence functional processes in litter decomposition and macroinvertebrate communities that, in turn, affect reproductive effort and success of waterfowl. Lilie and Evrard (1994) found that large, dense stands of submersed vegetation of Wisconsin wetlands offered greater shelter for macroinvertebrates from predation by birds, hence also had a higher biomass of invertebrates compared to less dense stands.

Less dense macrophyte stands provide only limited refuge for macroinvertebrates, hence better opportunities for feeding by birds and fish on them. They found that macroinvertebrate densities below 5000 individuals m^{-2} resulted in low densities of breeding pairs of blue-winged teal (*Anas discors*), but this had no effect on mallard (*Anas platyrhynchos*) pair densities.

A number of studies has shown a direct effect of fish on benthic invertebrates, and in this way competition between fish and waterfowl. Winfield and Winfield (1994) provided further evidence in their study of interactions between fish populations and overwintering tufted duck (*Aythya fuligula*) of Lough Neagh, Northern Ireland.

They leaned on previous studies, which found that such competitive interactions occur especially during the waterfowl breeding season when their protein requirements are high. Breeding hens utilise invertebrates as a source of proteins for egg-laying, and ducklings depend heavily on the consumption of invertebrates as a source of protein during their first few weeks of life. During such periods in life of the water birds, competition for food by fish may determine whether some water birds will select a particular water body for nesting and breeding. Erikson (1979) found that the distribution of fledged goldeneye (*Bucepala clangula*) was negatively associated with the presence of fish, mainly roach and perch. Giles *et al.* (1990) suggested that significant diet overlap exists between perch and ducklings of mallard (*Anas platyrhynchos*) and tufted duck.

In the above interactions, the observed or postulated competition resulted in waterfowl being out-competed by fish (Winfield and Winfield, 1994). In Lough Neagh, the largest lake in the UK, molluscs (especially *Lymnaea peregra* and *Valvata piscinalis*) were the dominant prey in the diet of tufted duck, followed by chironomid larvae. Molluscs also formed a major component of the diet of roach, while the diet of eel was overwhelmingly dominated by chironomid larvae. Tufted ducks are primarily carnivorous with molluscs dominating the diet and chironomid larvae typically assuming primary importance only when the former are unavailable.

Interestingly, in other localities, roach often feed predominantly on macrophytes or filamentous algae (see review by Lammens and Hoogenboezen, 1991) and such a plant-dominated diet was found by Giles *et al.* (1990) to minimise the overlap between roach and waterfowl in a macrophyte-rich lake.

The same authors also found that consumption of cladoceran zooplankton by adult roach minimised diet overflap between roach and waterfowl in a macrophyte-poor lake.

In Lough Neagh it is likely that the benthivorous roach population increase during the winter may result in a decline in the overwintering numbers of tufted duck (Winfield and Winfield, 1994). This further reconfirms the observations by Andersson (1981) on increasingly eutrophic lakes in southern Sweden, where declines in waterfowl were probably caused by competition for benthic food resources with increasing cyprinid populations, principally roach and bream (*Abramis brama*).

Food competition between a freshwater fish community and tufted duck was further confirmed through an experimental removal fish from a quarry lake (17 ha) in England (Giles, 1994). The fish community was dominated by adult bream and roach, with numerous small perch and abundant pike (*Esox lucius*).

The fish employed several mechanisms in the suppression of macrophyte growth including uprooting by benthivorous species (bream, carp), browsing of germinating seedlings (roach), and probably increasing nutrient cycling favouring an algal dominated open water plant community. Prior to the fish removal this lake had very little submersed aquatic vegetation. After the fish removal, dense beds of several species of submersed macrophytes grew over the whole lake bed. The post fish-removal increases in snail biomass have probably been central to the increased brood size at fledging recorded in this lake (Giles, 1994). The numbers of herbivorous wintering wildfowl (mute swan, *Cygnus olor*, coot, *Fulica atra*, and gadwall, *Anas strepera*) using the lake have increased dramatically following fish removal and the consequent surges in submersed aquatic plants. The probable effects of fish on aquatic birds in the quarry lake in England are shown in Table below.

Table: Probable effects of fish on gravel pit aquatic bird species and mechanisms of action. (From Giles, 1994)

Fish species	Diet	Birds affected (and mechanisms)
Bream, carp, tench	Chironomid larvae, molluscs	All ducklings via food competition. Herbivorous waterfowl via depressed submersed plant growth
Roach	Zooplankton, algae, plant seedlings	Tufted ducklings via suspension of plant growth and reduced snail production
Perch	Chironomid pupae, crustaceans	All ducklings via food competition
Pike reduction	Fish, ducklings	Predation of ducklings by small pike but also in food competition from insectivorous fish via predation by pike of all sizes

In Florida lakes, Hoyer and Canfield (1994) found that birds do not significantly impact the trophic status of the lakes under natural conditions, but lake trophic status is a major factor influencing bird abundance and species richness on lakes. They found no significant correlation between annual average abundance and species richness and macrophyte abundance. Bird abundance and species richness remain relatively stable as macrophyte abundance increases, but birds that use open-water habitats (e.g. cormorant *Phalacrocorax auritus*) are replaced by species that use macrophyte communities (e.g. duck *Aythya collaris*). On the other hand, Lodge (1991), referring to literature data, pointed out that waterfowl influence submersed and emergent plants locally and seasonally: snow geese (*Chen caerulescens*) substantially reduce macrophyte (predominantly emergent and wetland plants) biomass on their summer breeding grounds, at migration stop-over sites, and on their wintering grounds. Less gregarious waterfowl, e.g. breeding ducks and swans during the macrophyte growing season, also substantially reduce plants, perhaps over larger areas than colonially nesting geese.

The presence of black swans (*Cygnus atratus*) is closely bound with the presence of aquatic macrophytes. Black swans, introduced to New Zealand from Australia, are exclusively herbivorous. A shallow lake (Hawksbury Lagoon) benthic vegetation was for some time fully dominated by thick mats of filamentous green algae *Spirogyra, Rhizoclonium* and *Enteromorpha intestinalis,* with charophyte *Nitella hookeri* also dominating the biomass at times. The only angiosperm recorded was *Ruppia* (Mitchell and Wass, 1996). After a 2–3 year period of macrophyte and benthic filamentous algal dominance, the lake switched to phytoplankton dominance. The reason for this was not, however, the impact of swans through direct feeding on benthic algae, but rather the change in the depth penetration of light, which occurred as a combination of phytoplankton and other suspended matter increase. Under such conditions, the additional effect of black swans grazing might well be critical and lead to the decline in aquatic macrophytes (Mitchell and Wass, 1996).

The same authors also assessed the impact of the faecal loading rates to nutrient dynamics. The data suggest that waterfowl faeces have low ratios of nitrogen to phosphorus, so that in lakes where their contribution is large, the phytoplankton will tend to be dominated by cyanobacteria. However, in Hawksbury Lagoon black swans faecal contribution to nutrient dynamics was slight (Mitchell and Wass, 1995), less than half of that estimated for Canada geese on a small Michigan lake. Geese feed on pasture, thus bringing into water nutrients from external source, rather than recycling material like black swans, feeding on benthic algae. Mitchell and Wass (1996) constructed a conceptual model for the roles of black swans and other non-seasonal factors in decline in macrophyte biomass in shallow lakes.

In lakes where for whatever reason submersed aquatic macrophytes have disappeared, e.g. in eutrophicated water bodies, the regrowth of macrophytes may be delayed by waterfowl grazing (Lauridsen *et al.*, 1993). Kiorboe (1980) stated that grazing by coots has only a minimal effect on macrophyte growth because grazing often takes place outside the growing season of the plants. Coots, however, pull out whole plants and may influence the macrophyte composition and succession by removing especially plants still present during autumn and winter. In experiments, under the impact of waterfowl grazing, especially coot (*Fulica atra*), plants growing on mud were found to have been pulled out of the sediment and most of them had disappeared. A high number of of coots appeared on the Dutch Lake Zwemlust when perennial macrophytes became dominant after biomanipulation. Van Donk (1998) suggested that high grazing pressure by coots on *Elodea* and *Ceratophyllum* during the first years after biomanipulation probably promoted the rise of *Potamogeton berchtoldii*. This plant was not negatively affected by coots because this species forms nongrazable overwintering structures. The impact of sequential grazing on aquatic macrophytes by rudd and coots in Lake Zwemlust, as reflected in changes in the total biomass of macrophytes.

Information on the consumption of aquatic macrophytes by herbivorous birds is sparse: in Lake Zwemlust, Netherlands, the estimated annual consumption by coots was calculated as being between 30 and 1 200 kg. This compares with the consumption of 170 to 360 kg by the fish rudd (*Scardinius erythrophthalmus*) (Van Donk and Otte, 1996). Mitchell and Perrow (1998) suggested that birds may have a major impact on the dynamics of plant biomass in temperate waters only during a low macrophyte production - either early in the growth phase of seasonal species, when biomass is low, or in autumn, when seasonal plant growth has slowed or ceased. Arrival of migratory flocks may dramatically increase the grazing impact and lead to a substantial decline in plant biomass after seasonal growth has ended.

Olofson (1991) found significantly higher growth of fenced-in macrophytes (*Potamogeton crispus*) than of unfenced macrophytes in Lake Stigholm, Denmark. The fish grazing was of minor importance as most herbivorous fish were prior to that removed. He concluded this was the result of grazing on water plants by waterfowl. Jupp and Spence (1977) found that plants of *Potamogeton filiformis*, protected from waterfowl grazing, became up to 5.3 times larger than unprotected plants. Lauridsen *et al.* (1993) concluded that waterfowl will be able to delay the re-establishment of submersed macrophytes, especially in shallow parts, where waterfowl are able to reach the bottom when feeding. This might delay the recovery of eutrophic lakes. Concluding, one can say that birds typically consume only a small fraction of the annual aquatic macrophyte production. Their major influence appears to derive not from what they consume but from the loss of the future growth potential of plants, compounded through the growth period. The macrophyte biomass parameters that are relevant to zooplankton or fish may not always be so for birds, as birds are highly mobile and sampling in - say - two or four weeks interval is quite inappropriate. Birds mobility means that the option of emigration to seek a better habitat is always available (Mitchell and Perrow, 1998). Where large generalist herbivores such as aquatic birds continue to increase, they may become an impediment to managing lakes for the clearwater state: a small increase in herbivory by such birds could tip the balance toward the turbid phytoplankton state (Lodge *et al.*, 1998), and in turn this would impact on fish.

Manatee (*Trichechus spp*)

In Florida and Guyana, *Trichechus manatus* is known to consume 36 genera of macrophytes, but not water hyacinth when other plants are available. It was reported that they efficiently clear canals when present in sufficient density. Various ponds and canals in Guyana have been kept clear of aquatic weeds by *Trichechus inunguis* for many years (Anon., 1974). This species, which inhabits the tropical rivers of south America, feeds at any level from bottom to surface and is able to feed also on the floating vegetation. Their preference is for succulent aquatic macrophytes, but they will consume almost any aquatic plant (Ronald *et al.*, 1978). They prefer submersed to floating, and floating to rooted-emergent plants. In the Chagres River in Panama weed control using the manatee was unsuccessful (Schuytema, 1977, and references therein). Some authors pointed out some of the limitations and difficulties in attempting to utilize this species. While it is a voracious feeder it tends to graze selectively and, except in situations where it can be confined to a small area and thus forced to feed on all plants, it will bypass thick stands of one species to feed on another which it prefers. This

factor, coupled with its scarcity, its low reproductive rate and the high cost of managing it, suggests that it is unlikely to be widely used in weed-control programmes.

Estimates for wet weight food consumption by captive manatees (*T. manatus*) range from 5 to 11 % of body weight per day (Best, 1981). Etheridge *et al.* (1985) made a detailed field study of manatees of the Crystal River, Florida, where they determined that the longest animal feeding in a 24 hour period was 6 hours and 10 minutes, with an average of 5 hours per day, an equivalent of approximately 7.1% of their own body weight. The introduced hydrilla was the principal food of manatees there. The authors, using Haller and Shireman (1982) data on aquatic plant biomass for Kings Bay, where the plant covered 165 ha, concluded that 18 manatees per ha would be required just to maintain the hydrilla at a constant biomass by consuming daily productivity. This is nearly ten times as many manatees than are present in the area. They argue that previous assessments of efficiency of manatees in controlling aquatic plants might have been exagerated, unless the local plant productivity was much lower than that in Kings Bay, and conclude that manatees cannot be considered efficient control agents for aquatic macrophytes.

Trichechus senegalensis inhabits the west coast from Senegal to Angola in Africa, but nothing is known about its plant-feeding habits.

Nutria (*Myocaster coypus*)

Schuytema (1977) reviewed the impact of nutria on aquatic plants. This animal, often introduced for its pelt and meat, is able to control the emergent *Typha angustata* and *Phragmites australis* in ponds in Europe and Israel, and water grass (*Echinochloa*) in Africa (Cameroon). This it does quite efficiently, and where nutria is present close to common carp ponds it can significantly increase fish production by destroying the emergent vegetation. The burrowing habit of nutria is a negative aspect of its presence as it may lead to damage to pond embankments, canal banks and dams. In Kenya nutria (coypu), which was imported for fur-farming in 1950, escaped and arrived at Lake Naivasha from 1965 onwards, where, by the early 1970s, there was a large population (Harper et al., 1990). Coypu are generally blamed for the disappearance of water-lilies (*Nymphaea caerulea*) from Lake Naivasha, on which individuals were widely observed to be feeding. Water-lily shoots make up a major part of the coypu's diet (Gibson, 1973). Water-lilies began to disappear in the eastern part of the lake at around the same time that the crayfish *Procambarus clarkii* populations were high (4 ind. m^{-2}). The crayfish had a substantial effect upon the submersed vegetation, which was reported to be abundant in 1979 but had completely disappeared from the lake by 1983, the last year in which crayfish harvest exceeded 100 tons. It is likely that water-lilies disappeared under the combined grazing pressure of coypu and crayfish (Harper *et al.*, 1990).

In the Norfolk Broads, UK, the highest rate of decrease in young reed shoots was found when the population of coypu was very high (Boorman and Fuller, 1981).

Muskrat (*Ondatra zibethicus*)

In the Czech Republic muskrat consumed or used in lodge construction 9–14% of the annual net biomass production of *Typha latifolia* (Pelikan *et al.*, 1971). In a lake in northern Germany a population of muskrats consumed or damaged 0.27 ha of *Typha*, 0.15 ha *Phragmites australis*, 0.86 ha *Glyceria*, and 1.58 *Scirpus*. The damaged area of *Scirpus* and *Phragmites* did not recover even when the muskrats significantly decreased in number (Akkermann, 1975). In the USA muskrat in the Atchafalaya Bay, Louisiana, graze on submersed swards, for example *Eleocharis* (Fuller *et al.*, 1984).

AQUATIC MACROPHYTES, WATER QUALITY, AND FISH

Water Quality

Water bodies with medium and dense macrophyte cover are characterized by a low concentration of suspended sediments, hence high water transparency. Where aquatic macrophytes disappear, such as under the impact of eutrophication, water transparency is reduced. In lakes where macrophytes reappear after a period of absence, water transparency gradually improves with the increasing vegetation cover. The success of prey capture by fish which locate their prey visually, such as pike, depends on good water transparency. Pikeperch tolerates a lower water transparency than pike.

Several of the chemical parameters influenced by plants that have been observed to reach lethal limits for common freshwater fish are dissolved oxygen, acidity and ammonia. Frodge *et al.* (1987, 1995) measured dissolved oxygen and pH values in patches of *Brasina scherberi* in the northwest of the USA and found their values ranging from above (in open water) to below (in the patches of the macrophyte) lethal limits for rainbow trout (*Oncorhynchus mykiss*) within 10 m of each other in the same lake. Rainbow trout in cages positioned at 1 m depth below *Nymphaea odorata*, where dissolved oxygen concentration was consistently less than 4 mg L^{-1}, were found dead within 12 hours. Significant mortalities of largemouth bass occurred at both the surface and at 1 m depth in dense beds of the floating leaved macrophyte *Brasenia schreberi*, and in the bottom water of 2 m or more depth of both floating leafed and submersed species, where dissolved oxygen concentrations were less than 2 mg L^{-1}.

The authors suggested that high densities of aquatic plants can have significant detrimental localized effects on fish and reduce the amount of habitat available to fish within the system. Perna and Demilio (1986) found that dissolved oxygen concentrations in the Ticino River, Italy, with submersed aquatic plants, changed by up to 4.0 mg L^{-1} between day and night. A trout farm, receiving water fom this river, experienced difficulties as a result of the temporary low oxygen concentrations.

Dissolved oxygen content in water may substantially change between day and night when large stands of aquatic macrophytes are present. This may result in substantial diel movement of fish within the plant bed to avoid low dissolved oxygen concentrations in the morning and high pH in the afternoon (Killgore *et al.*, 1991).

In many Amazonian floodplain lakes dissolved oxygen concentrations are often very low and anoxia is frequent at least in deeper waters. Lago Camaleao in Brazil is situated about 15 km above the confluence of the Amazon River with Rio Negro. During low water, the lake is dry, apart from a small, muddy pool of 1–2 ha of 50–80 cm depth (Junk *et al.*, 1983) and the dry lake bottom is covered by terrestrial and semi-aquatic herbaceous plants, which produce considerable amount of organic material. When the water level rises, the lake becomes flooded, the terrestrial plants die and decompose, while aquatic plants develop in great quantities, with grasses being the most abundant among them.

When some of the aquatic plants start dying, about 60% of the lake is still covered by the grass *Echinochloa polystachia*, and floating plants *Eichhornia crassipes* and *Salvinia auriculata* become abundant. The highest levels of dissolved oxygen have been recorded in the lake when there are still large open water spaces, but a heavy oxygen deficit is present when there is a large-scale plant decomposition in progress. This may reduce the values at the water surface to 0.5 mg L^{-1}. An increase in dissolved oxygen concentration is noticed with the reduction in the macrophyte cover. Junk *et al.* (1983) describe the consequences of the low dissolved oxygen concentrations in this lake: the fish is forced to expend much energy and time in the search for oxygen, for instance coming to the surface to swallow air, or staying near the surface to use the oxygen-rich surface layer.

This may increase the danger of being caught by predators. Food diversity is considerably reduced because phytoplankton and zooplankton become scarce. The bottom fauna is lacking because of the lack of oxygen. The authors also list the advantages of the low dissolved oxygen concentrations among aquatic macrophytes of Lake Camaleao: interspecific competition for food is reduced, as only a small number of fish species utilizes the existing food sources; and there is a reduction in predators.

The presence of many juvenile fish under hypoxic conditions may be related to this factor. In Lake Camaleao, about 40 species of fish were able to live under extremely low oxygen conditions, 12 of which were air breathers, and 8 possessed other adaptations to the oxygen deficiency. For example, *Colossoma macropomum*, is known to be able to increase its haemoglobin content and erythrocyte counts associated with seasonal hypoxia. This species initiates aquatic surface respiration when dissolved oxygen level falls below 0.5 mg L^{-1} (Saint-Paul, 1984). Jedicke *et al.* (1989), who recorded diurnal changes in dissolved oxygen concentration beneath floating macrophytes in a varzea lake, showed that while the maximum concentrations exceed 12 mg L^{-1} in the afternoon hours, in the early morning hours the water is virtually deoxygenated.

Junk (1973) measured vertical oxygen profiles in stands of *Leersia hexandra* in Lago dos Passarinhos (25 km from Manaus, Brasil). The profile was similar to that measured in dense stands of *Paspalum repens* of varzea: high dissolved oxygen concentrations at the surface, rapid decrease into the roots, and a slight increase at the lower margin caused by an inflow of oxygenated water from open areas. Within 20 cm depth the dissolved oxygen concentration dropped to less than 0.5 mg L^{-1}, and H_2S sometimes formed at a depth of 30 to 140 cm. In a dense, floating island in Lago Paru, 10 m inside the mat of *Paspalum repens*, the dissolved oxygen concentration dropped from 6–8 mg L^{-1} at the surface to almost nil in a depth of 20 cm, to increase to about 3 mg L^{-1} at 30 cm and deeper. Junk (1973) also noted that low conductivity of water is not a limiting factor for faunal abundance in the floating meadows of the varzea, until it falls below 30 iS_{20}.

In the reservoir Bung Borapet in Thailand, Junk (1977) recorded dissolved oxygen profiles underneath stands of several aquatic macrophytes. The lowest concentrations were measured 10 m inside the stand of *Coix aquatica*, where there was no oxygen present already at 10 cm depth. Very low concentrations were also measured underneath *Leersia hexandra* and *Ischne globosa*, while underneath water hyacinth the lowest dissolved oxygen concentration, measured above the bottom, was 2.5 mg L^{-1}. All four plants form floating islands which drift around the reservoir.

Hypoxia is widespread in tropical fresh waters, particularly in floodplain pools and permanent swamps. It is prevalent in extensive papyrus swamplands (Petr, 1973), underneath sudd (Junk, 1977; Junk and Robertson, 1997), in deeper waters of the newly established reservoirs (Viner, 1970), and it may pose a formidable barrier to the dispersion of many fish groups. While some non-air breathing fish may occasionally penetrate in the almost deoxygenated papyrus swamps of East Africa, the fish inhabitants there are primarily air breathers such as *Protopterus aethiopicus*, *Polypterus bichir*, *Clarias lazera*, *Ctenopoma muriei* (Beadle, 1981). In a small cyprinid *Barbus neumayeri*, not an air-breathing fish, the respiratory mode affected its dispersal through hypoxic papyrus swamps (Chapman and Liem, 1995).

This fish, which feeds on insect larvae and aquatic plants, was found to inhabit swamp waters with dissolved oxygen concentrations between 0.34 and 1.24 mg L^{-1}, but always in places with at least 0.6 m^2 of open water surface where it has access to the water surface and higher oxygen. In response to severe hypoxia it uses aquatic surface respiration at the air-water interface, where diffusion produces a very thin layer of well oxygenated water, despite oxygen scarcity in the rest of the water column. Chapman and Liem (op.cit.) also found significant differences in the gill morphology between *B. neumayeri* from the papyrus swamp and those from the main river, thus affirming the need for a morphological adaptation, as well as having access to the open water within the swamp.

But high water quality is not a necessary prerequisite for optimal fish production, unless fishery managers are to satisfy a demand for fish species that can exist only in oligotrophic waters, such as salmonids and coregonids. With most cyprinids, the higher the trophic level, the higher fish production - to a certain level. This approach has been used in common carp ponds in Europe and Asia. On the other hand, where sport fishery targets fish species in clean rivers, any deterioration of water quality results in the reduction or loss of such species.

Eutrophication and Aquatic Macrophytes

Eutrophication has become a problem in many countries. De Nie (1987) has briefly reviewed long-term changes in aquatic macrophytes of selected water bodies under the impact of eutrophication in Finland, Germany, the Netherlands, Poland, Switzerland, UK, and the USA. For example in some small water bodies in the Netherlands, the area covered by submersed vegetation declined by a factor 6. In the polish lake Mikolajskie (460 ha) over the period 1963–1980 the decline in the biomass of aquatic macrophytes was 16.6% per year, which equalled to 94% over the entire period. In five eutrophicated lakes of Finland the submersed vegetation had greatly decreased or became restricted to small areas.

Also in Germany the eutrophication influenced especially the submersed macrophytes, and in Lake Constance the depth of occurrence of the Characeae changed from a maximum of 25 m in the 1950s to 5–7 m in 1978. In shallow lakes of the Norfolk Broadlands, palaeolimnological study has shown that starting with the 19th century Strumpshaw Broad was rich in Characeae and epiphytes, with productive fisheries and rich in wildlife (Moss, 1990). In the beginning of the 20th century the Characeae were replaced by taller macrophytes, such as

Potamogeton pectinatus, *Myriophyllum spicatum* and *Ceratophyllum demersum*. During that period the eutrophication started by sewage effluents from the nearby city and cultivation and fertilization of the surrounding land. The fisheries was still productive. This was followed by changes towards the middle of the 20th century, and especially afterwards: by the early 1970s aquatic macrophytes disappeared and phytoplankton started to predominate. The fisheries became less productive and there was an increase in the erosion of the banks and in sedimentation.

Inputs of nitrogen and phosphorus especially in shallow freshwater bodies have led to considerable changes in the aquatic community structure, enhancing phytoplankton production, which in turn has caused suppression or complete disappearance of aquatic submersed macrophytes. Such changes have also meant changes in aquatic invertebrate and fish species composition and biomass. Based on their experiments inducing artificially eutrophication using high nitrate treatment, Daldorphy and Thomas (1991) identified three possible mechanisms which may be responsible for the rapid transition to phytoplankton dominance:

(i) competition by phytoplankton for resources such as light and nutrients, where high phytoplankton density attenuates the photosynthetically active radiation impinging on macrophytes;

(ii) inhibition of macrophyte growth by toxic factors of algal and bacterial origin. During algal blooms, considerable quantities of organic matter may accumulate in the sediment. The addition of such material greatly reduces the growth of submersed macrophytes, possibly due to the phytotoxic organic acids released during decomposition;

(iii) inhibition of macrophytes by epiphytic algae.

Biomanipulation

The term "biomanipulation" was used by Shapiro *et al.* (1975) to include lake improvement procedures that alter the food web to favour grazing on algae by zooplankton, or that eliminate fish species that recycle nutrients. Cooke *et al.* (1988) expanded this term to the elimination of certain species or restructuring of the fish community to favour the dominance of piscivorous instead of planktivorous fish. In broad terms, biomanipulation is the management of aquatic communities by controlling natural populations of organisms to produce desired conditions. Lazzaro (1997) refers to biomanipulation as a set of deliberate modifications of certain key communities, and/or their habitats, to improve water quality/clarity and/or (less often) fisheries characteristics. Van Donk (1998) uses this term in a sense of man-made modification of fish populations.

The topic of biomanipulation has been widely covered in a symposium on biomanipulation, held in 1990 (Gulati *et al.*, 1990). A Guide to the Restoration of Nutrient-enriched Shallow Lakes (Moss *et al.*, 1996) also includes a chapter on Biomanipulation, and the topic is a subject of a special FAO fisheries publication, to be published shortly. Numerous papers exist on this topic, especially discussing the long-term observations on Dutch shallow lakes, especially Lake Zwemlust. For the purpose of the present publication only the biomanipulation component of using fish to achieve improvement of water quality, and hence the return or enhancement of aquatic macrophytes, is discussed.

The principal objective of biomanipulation in shallow lakes is to generate a period of clear water of sufficient length to allow macrophytes to establish (Perrow *et al.*, 1997). Both a relative increase in the piscivorous fish stock and a reduction in nutrient levels are thought to be important in stabilising the system in the long term. With increase in lake trophy, the area occupied by and biomass of submersed macrophytes decreases, and in hypertrophic lakes aquatic plants are scarce or absent (Ozimek *et al.*, 1990). Disappearance of macrophytes leads to other changes in the aquatic communities.

Highly eutrophic to hypertrophic lakes have a characteristically high phytoplankton production and in Europe large stocks of plankton-feeding fish, such as bream (*Abramis brama*) and roach (*Rutilus rutilus*). To improve the water quality, as well as to establish a higher fish species diversity, including fish of fishery significance, a technique of biomanipulation has been developed. Biomanipulation aims at decreasing the phytoplankton biomass, therefore increasing water transparency, which in turn leads to re-establishment of aquatic macrophytes. The macrophytes provide then a substrate for periphytic fauna and lead also to an increase in bottom-dwelling invertebrates, both of which diversify the source of food for fish.

In a shallow and small Lake Zwemlust in the Netherlands, biomanipulation involved complete removal of all planktivorous and benthivorous fish species (ca. 1000 kg ha^{-1}), introduction of a new simple fish community (pike and rudd) as well as of "seedlings" of *Chara globularis* and roots of *Nuphar lutea*. While before the manipulation no submersed vegetation was present in the lake, the second year after the manipulation about 50% of the lake bottom was covered by macrophytes dominated by *Elodea nuttallii*, increasing to 80% in the third year (Van Donk *et al.*, 1990). The expansion of macrophytes was possible due to the drastic reduction in phytoplankton, especially cyanobacteria. Other changes resulting from the biomanipulation was development of strong populations of zooplankton and benthic/ epiphytic macrofauna, especially snails *Lymnaea peregra*, and a pike/rudd fish community. In summer and autumn the abundance of macrophytes and filamentous algae led to nitrogen limitation of the phytoplankton. The persistence of clear water in the summer after the biomanipulation was believed to be caused by macrophytes, rather then by zooplankton grazing on algae.

Van Donk (1998) followed the changes in Lake Zwemlust for 10 years and noted that after the initial biomanipulation the lake shifted several times between the turbid and the clearwater state. In years 4–5 *Ceratophyllum demersum* became the dominant macrophyte. Both *Elodea* and *Ceratophyllum* compete strongly with phytoplankton for nutrients, especially nitrogen. In the next three years, these plants became nearly absent, and *Potamogeton berchtoldii* became the dominant species in spring. However, periphyton growth on this plant led to its early collapse, which was followed by an increase in phytoplankton and the lake became turbid during late summer and autumn. In 1995 *E. nuttallii* reappeared and became again the dominant macrophyte. Next year, it became scarce and *Potamogeton* abounded, and the lake shifted again to the turbid state (Van Donk, 1998).

She concluded that the presence of submersed macrophytes seems to be essential in keeping this lake clear. However, she added that herbivory, especially by coots (*Fulica atra*), and to a lesser extent by rudd, was probably an important factor in triggering the shift from perennials (i.e. *Elodea* and *Ceratophyllum*) to non-perennials (*Potamogeton*). The number of larger rudd (only larger rudd are herbivorous) was quite low until 1990 but increased in 1991 to 297 kg ha^{-1}, to stabilize at 200–300 kg ha^{-1} during 1992–1996. As rudd feeds only when water temperature exceed 16°C, and avoids *Ceratophyllum*, Van Donk (1998) suggested that it had much lower impact on total macrophyte biomass than coot.

The biomanipulation of Lake Zwemlust by re-establishment of macrophytes and filamentous algae has created a habitat for periphytic fauna and changed the environmental and trophic conditions for fauna inhabiting the bottom sediments. This was achieved mainly by changing the physical properties of bottom sediments and the water, especially water transparency and the dissolved oxygen concentration. Zoobenthos achieved a higher biomass than phytofauna, except in submersed plants, especially *Elodea nuttallii*, which occupied the majority of the bottom surface (Kornijow *et al.* (1990). Filamentous algae (e.g. *Mougeotia* sp. and *Oedogonium* sp.) were readily eaten live by chironomid larvae and during the algal decay in autumn, when sedimenting, also by the benthic *Lymnaea peregra*, *Asellus meridianus* and *Gammarus pulex*.

In order to improve lake water quality in Lake Vaeng in Denmark, a total of 2.5 tons of bream and roach were removed, reducing the biomass of planktivorous/benthivorous fish by approximately 50% (Sondergaard *et al.*, 1990). Again, there was a change from cyanobacteria and small diatoms to a dominance of large diatoms, larger greens and cryptophytes, change in zooplankton from the dominance of rotifers to large cladocerans, and increase in water transparency and dissolved oxygen content. Submersed macrophytes, mainly *Potamogeton crispus* and *Elodea canadensis*, increased in abundance. In the following years, *Elodea* coverage fluctuated. This was paralleled by the fluctuation of aquatic birds coot and mute swans, with their high numbers recorded only in years with high macrophyte density (Sondergaard *et al.*, 1998). A similar parallel between the density of aquatic macrophytes and coot were also observed on another Danish lake Stigsholm. The grazing intensity in Lake Vaeng is suspected to inhibit colonization of the lake by aquatic plants. In both Danish lakes the results of the Sondergaard *et al.* (1998) studies suggest that waterfowl may suppress macrophyte biomass and development in shallow lakes, and thus have a negative impact on some lake fish.

In a large eutrophic lake in Minnesota, USA (Lake Christina) the water was highly turbid prior to biomanipulation (summer transparency less than 30 cm), virtually devoid of macrophytes, with large stocks of planktivorous fish bluegill (*Lepomis macrochirus*), small yellow perch (*Perca flavescens*), and big-mouth buffalo (*Ictiobus cyprinellus*). Benthivorous black bullheads (*Ictalurus melas*) were also abundant (Hanson and Butler, 1990). The zooplankton community was typical of those subjected to intense planktivory by fish, being dominated by small-bodied species of cladocerans, by *Cyclops bicuspidatus thomasi*, and several rotifers. The fish removal by rotenoning, which led to virtually all fish being killed, followed by restocking, resulted in a much improved water transparency and dramatic changes in zooplankton, phytoplankton, and in submersed vegetation (*Chara, Najas, Ruppia* and *Myriophyllum*) revival. Unlike other lakes, in Lake Christina the post-manipulation increases in water clarity and submersed macrophytes were caused by the increases in zooplankton grazing, with the peak daphnid abundance in late May-June. These months are the critical growth period for submersed macrophytes.

In Europe, pike and pikeperch have been used to suppress the small zooplanktivorous fish such as bream, roach and perch. In fully overgrown lakes, a high cyprinid (<15 cm) production as high as ca. 300 kg ha^{-1} can be consumed by a high pike biomass (100 kg ha^{-1}) (Grimm and Backx, 1990). Such a high production of planktivorous cyprinids is associated with a total phosphorus concentration of ca 400 ìg P L^{-1}. At a higher biomass of cyprinids the large-bodied zooplankton disappears, phytoplankton increases, causing a decrease in the macrophytes area, and leading to the dominance of *Potamogeton pectinatus* and plants with floating leaves. This in turn leads to changes in fish stocks, with the pike-tench associations being replaced by bream-pikeperch populations. There is an increase in benthic-feeding fish, which in turn increases the internal loading and turbidity. Grimm and Backx (1990) concluded that in shallow eutrophic lakes of northern Europe aquatic vegetation and northern pike are effective tools to maintain water quality, but these are limited by the maximum nutrient concentration aquatic vegetation can sustain. Pikeperch is more efficient then pike in turbid water, therefore, in a more eutrophicated environment (Moss *et al.*, 1996). Addition of piscivores, however, will not have a lasting effect unless they are stocked every year. Therefore, removal of the fish stocks and maintaining them at very low levels for a number of years gives better effect. The available data suggest that in phytoplankton-dominated lakes, the total fish biomass will be between 200 and 500 kg ha^{-1}. Where only half, or somewhat more of thek biomass has been removed, there has been no real improvement (Moss *et al.*, 1996).

Bronmark and Weisner (1992) highlighted the indirect effects of the fish community on submersed vegetation in shallow eutrophicated lakes: eutrophication and the accompanying phytoplankton and periphyton dominance shade off the macrophytes leading to their demise. This leads to decline of piscivorous fish species, and an increase in phytoplankton feeding fish. Return from the phytoplankton-dominated state depends on an increase in piscivore stocks, either through restocking or growth of the small surviving piscivores. The resulting decrease in phytoplankton biomass and increase in light availability will promote the reestablishment of submersed macrophytes. In reality, this should be accompanied by substantial reduction of the planktivorous fish stocks, to achieve good results (Moss *et al.*, 1996; Van Donk and Otte, 1996).

A study of lakes in Florida has shown that leaving a small fringe of vegetation around a lake for the purpose of water quality improvement will have little or no effect on the lake's trophic status (Canfield and Hoyer, 1992). Significantly reducing macrophyte coverage of a lake, for example from 60% to 20% or from 40% to 0% will cause major and observable water quality (water clarity) changes. Conversely, the cover must be raised to 30% to 50% before significant improvement in water clarity will be observable.

The general contention of biomanipulation is that removal of fish from shallow lakes may invoke "cascading impacts" that enhance water clarity, stimulate macrophytes, and establish a new steady state. Recent, large-scale biomanipulations have made it possible to update earlier recommendations regarding when, where, and how biomanipulation should be performed. Hansson *et al.* (1998) formulated more applicable recommendations as follows:

(1) the reduction in the biomass of planktivorous fish should be 75% or more;

(2) the fish reduction should be performed efficiently and rapidly (within 1–3 years);

(3) efforts should be made to reduce the number of benthic feeding fish;

(4) the recruitment of young-of-the-year fish should be reduced;

(5) the conditions for establishment of submersed macrophytes should be improved; and

(6) the external input of nutrients (phosphorus and nitrogen) should be reduced as much as possible before the biomanipulation.

Recent biomanipulations have shown that, correctly performed, the method also achieves results in large, relatively deep and eutrophic lakes, at least within 5 years. Moss *et al.* (1996) provide decision trees for biomanipulation and for aquatic macrophytes establishment.

Lazzaro (1997) addressed the question whether trophic cascades and 'classical' biomanipulation apply to the tropics. The 'trophic cascade hypothesis', developed for temperate systems, does not seem to be a good candidate for tropical lakes, for several reasons, among which is its typical linear structure from piscivores '! zooplanktivores '! zooplankton '! phytoplankton, which disregards the widespread omnivory over size-selective zooplanktivory per se. In direct reference to fish Lazarro mentions the seasonality of spawning in temperate waters compared to the spawning of tropical fish, which may spawn at different times, therefore juvenile fish maintaining a permanent pressure on zooplankton.

Filtering omnivores (not visual zooplanktivores) often dominate the planktivore community in tropical waters, without being limited by the presence/absence of zooplankton, nor controlled by fish predators as because of their fast growth they soon reach size larger than the predator's mouth gape. Tropical waters have smaller, 'sedentary' carnivorous fish, unlike temperate waters where large-sized piscivores dominate. Macrophytes in tropical waters grow faster and all the year. They can provide serious management problems, as their proliferation is not controlled in eutrophic systems, and they represent a highly variable and unstable sink or source of nutrients. Concluding, Lazzaro says that the above characteristics of tropical lakes and reservoirs weaken the top-down links between piscivores and planktivores, and between zooplankton and phytoplankton. Because of their weakness, there is little chance for a successful biomanipulation.

Aquatic Macrophytes as Wastewater Treatment System, and Fish

Lakes and ponds bordered by swamps and marginal vegetation are widely exploited for municipal, industrial and agricultural wastewater treatment. Aquatic macrophytes may also be used for the removal of aquaculture effluents. This assists in maintaining the water quality in the water body concerned at an acceptable standard for fish in natural water bodies and reservoirs.

In the 1970s the treated effluent from the often overloaded Kampala (Uganda) main sewage works entered the papyrus along the north shore of Lake Victoria, and its distinct chemical signature was completely erased after it passed through a kilometer of swamp into the lake (UNDP/WHO, 1970). The use of the large emergent species *Schoenoplectus lacustris* and *Phragmites australis* in Europe, and *Eichhornia crassipes* in USA, for treating industrial and urban effluents has been mentioned by Finlayson and Mitchell (1982). Water hyacinth prefers eutrophic conditions, and an increase of nitrogen level from one to 25 mg L^{-1} in water increases their dry weight and plant number linearly (Chadwick and Obeid, 1966). Dense submersed macrophyte beds with their associated epiphytic algae are considered in some areas to be useful nutrient filters for absorption of nitrogen and phosphorus (Howard-Williams, 1981). Use of constructed wetlands for surface-flow effluent treatment in the USA and Canada was initiated by a series of pilot-scale trials at Arcata in northern California, Gustine in southern California, Orlando and Lakeland in Florida, Listowel and Port Ferry in Ontario, Canada (Gaerheart, 1992). In the Netherlands constructed wetlands are applied to intermittent flows such as campsites, or sewage disposal from small communities.

Aquatic macrophytes have been also successfully used for removal of nutrients originating from fish production. In the experimental use of macrophytes for removal of nutrients originating from fish production

(*Oreochromis aureus*) tanks, water hyacinth was the fastest growing plant in the wastewater treatment tank and removed the greatest amount of nitrogen per square metre of growing area (Rakocy and Allison, 1981). The submersed *Egeria densa* was also an efficient plant in nitrogen removal, while *Vallisneria* sp. grew slowest and removed the lowest amount of nitrogen. The combined plant population, consisting of two floating and two submersed plants, removed 12 to 15.5% of the waste nitrogen in the wastewater treatment tank. More information on the removal of nutrients from fish culture using aquatic plants is the subject of another FAO publication.

Wastewater discharge to wetlands is often expressed in litres per square meter per day, which is termed the hydraulic loading rate (HLR). In a study carried out by Schwartz and Boyd (1995) a free water surface wetland was constructed adjacent to a commercial fish pond used to raise channel catfish (*Ictalurus punctatus*), and its efficiency in removing potential pollutants from pond water was evaluated. Two experimental wetlands were constructed in series, with the first receiving the effluent directly from the pond, being planted with *Scirpus californicus* and *Zizaniopsis miliacea*, and the other - receiving the effluent from the first one - planted with *Panicum hemitomon*. The HLR were 77–91 L m^{-2} day^{-1}. The concentrations of potential pollutants were much lower in the effluent from the wetlands than in the inflows from ponds, with the following reductions in concentrations recorded: total ammonia nitrogen, 1–81%; nitrite-nitrogen, 43–98%; nitrate-nitrogen, 51–75%; total Kjeldahl nitrogen, 45–61%; total phosphorus, 59–84%; biochemical oxygend demand, 37–67%; suspended solids, 75–87%; volatile suspended solids, 68–91%; and settleable solids, 57–100%.

Overall performance of the wetland was best when operated with a 4-day hydraulic residence time (HRT) in the vegetative season, but good removal of potential pollutants was achieved for shorter HRTs and when vegetation was dormant. The disadvantage of wetlands for treating aquaculture pond wastes is the large amount of space necessary to provide an adequate HRT. Wetlands also could be used for treating overflow from ponds after rains. The experiments have also shown that *S. californicus* and *Z. miliacea* provided much thicker stands than *P. hemitomon*, with the *S. californicus* being the fastest growing plant. Hence the use of *S. californicus* alone or together with *Z. miliacea* would provide for quicker plant establishment and higher standing crop in wetlands than the combination used in the study.

Wetlands have also been used on a limited scale for the removal of heavy metals. However, total elimination of toxic materials, such as heavy metals, may not be achieved. Copper mining in the foothills of the Ruwenzori Mountains, Uganda, has been a source of heavy metals from the leaching dumps and tailings, with the toxic metals reaching a swamp along the northern shore of Lake George. The succession of plants ends at the lakeward edge with floating rafts of papyrus. In spite of some removal of the heavy metals, fish caught at the swamp edge have distinctly higher concentrations of copper in their liver compared with fish from the outflow from Lake George (Kazinga Channel) and those from the nearby Lake Edward. As no elevated concentrations of heavy metals in the fringing aquatic macrophytes have been found, Denny *et al.* (1995) suggested that a proportion of metal-enriched water and sediments must pass through the swamp. The highest concentrations are in the phytoplankton and the floating *Pistia stratiotes* near the swamp fringe. Elevated copper concentrations were also found elevated in zooplankton and in the livers of juvenile tilapias (*O. leucostictus*), which fed on it.

To learn more on the use of natural and constructed wetlands for wastewater treatment in Europe, the reader is advised to consult other publications, such as Hamer (1990), Moshiri (1994), Vymazal *et al.* (1998) and Vymazal (1999). For constructed wetlands for the treatment of landfill leachates see the publication by Mulamoottil *et al.* (1998).

Intensification of Fish Production in Fish Ponds, and Aquatic Macrophytes

Fish ponds in southern Bohemia (Czech Republic) are a classic example of the impact of human activities on aquatic environment, as manifested by changes in aquatic plants and fish production in these water bodies. The fish ponds, constructed in the 14th to 16th centuries on poor soils, had initially oligotrophic to dystrophic character, as their water originated from large peat bogs. Such ponds had a high aquatic macrophyte species diversity, high water transparency, and well developed periphyton and benthos, grazed and consumed by molluscs, insect larvae and fish. The dissolved oxygen concentration was high and almost constant, about 100 % saturation (Pokorny *et al.*, 1990).

Fish production in such fish ponds was very low. Intensification of agriculture in pond catchments, which started in the 1930s, the input of manure and fertilizers directly into the ponds, and in some cases also provision of fish feed, have led to the establishment of eutrophic and even hypertrophic conditions. This was accompanied by increasing fish production, but also by increasing fluctuations in pH and dissolved oxygen concentrations (Pokorny *et al.*, 1992). In nutrient-rich ponds well-developed periphyton started shading off the macrophytes leading to their decline. Within twenty years the vegetation typical of oligo-mesotrophic waters disappeared, being replaced in the inshore areas by Potamogeton perfoliatus, tolerating the eutrophicated environment.

In summer there was a strong development of phytoplankton, which was dominated by chlorococcal algae, and this added to the shading effect, further reducing water transparency. Such ponds are suitable for carp culture, with fish production reaching up to 1 000 kg ha^{-1}, but they have low fish species diversity, especially being short of predators. With virtually no permanent aquatic macrophytes that would help to stabilise the aquatic community, pond managers have to keep the fish pond ecosystem in a steady state ensuring high fish production on the one hand and reasonable water quality on the other, as fish ponds in southern Bohemia serve also other functions, such as recreation.

The Czech fish ponds are an example of a situation where macrophytes are considered redundant, as these ponds are directed toward fish production along the shortest possible food chain, i.e. phytoplankton-zooplankton-fish (Pokorny *et al.*, 1990). In eutrophic/hypertrophic fish ponds the major threat to the fish stocks comes from the highly variable and sometimes hazardous environment. Such conditions, common in densely stocked and fertilised or manured fish ponds, can be tolerated by a few fish species such as the large-sized common carp, but less so by fry or fingerlings. Pokorny *et al.* (1990) suggest that in order to improve the ecological conditions for fish, and especially for the fry, it would be useful to reconsider the role of aquatic macrophytes as a component stabilising the fish ecosystem.

The main stabilising properties of aquatic macrophytes in fish ponds are their relatively slow biomass turnover and high net production, their relatively high uptake of nutrients which remain fixed in their biomass or detritus for some time and may be harvested when required, and the provision of shelter for young fish. Phytophilous aquatic macroinvertebrate fauna is an important component of the fish diet, and some waterfowl feed intensively on some macrophytes or their parts. The detritus accumulating in macrophyte stands binds mineral nutrients and organic matter at times of their abundance in the fish ponds and releases them gradually according to the dissolved oxygen pattern and diffusion in the sediments. The detritus serves as a source of food for benthos and also for filtrating zooplankton when algal production is low. Detritus is also the most important substrate for microbial decomposers.

AQUATIC MACROPHYTES AS A LINK IN THE FOOD CHAIN

Effect of Fish Grazing on Nutrient Release

Herbivorous fish have a limited ability to convert plant material into animal tissue and to remove excess nutrients from the system. But being relatively efficient grazers and inefficient assimilators, they may play a significant part in initiating the internal, biologically mobilized nutrient source, or accelerate their recirculation (Prejs, 1984). Thus, if with advancing eutrophication, the rate of macrophyte consumption increases due to the increase in number of herbivores, a new part of their biomass may finally be converted into additional algal biomass - the main troublemaker in lakes of the temperate zone.

Considerable information is available on the impact of grass carp grazing on water quality. Grass carp partially digests its food and the faeces are a large source of nutrients. Nutrients are readily utilised by phytoplankton, resulting in elevated algal biomass, and in some situations leading to algal blooms. Bettoli *et al.* (1993) monitored the changes in chlorophyll *a* concentration prior to and after the introduction of grass carp in Lake Conroe.

Not always the primary production changes under the impact of grass carp grazing on macrophytes. In Red Haw Lake, Iowa, where within three years the introduced grass carp reduced aquatic macrophytes from 2 438 g m^{-2} to 211 g m^{-2}, mean nitrites, nitrates, biological oxygen demand, and turbidity showed significant decreases, while alkalinity increased significantly (Mitzner, 1978). Mean concentrations of organic and inorganic phosphates gradually increased, but were not statistically different. During the first two years the average primary production was nearly identical, followed by a decrease in the third year. Grass carp consumed all major plant groups, with preference for *Najas* and *Potamogeton*. Lembi *et al.* (1978) found that as much as 54% of the phosphorus and 42% of the nitrogen released by consumption of plants by grass carp in Indiana ponds were incorporated into new fish tissue. Most of the phosphorus not taken up by the fish was sequestered into components other than water or phytoplankton. The same authors also noted that potassium concentrations in water appeared to be an excellent indicator of vegetation consumption by grass carp. The authors suggested that the impact of the grass carp on water quality should be minimal in temperate zone small water bodies if the fish is stocked early in the season when vegetation biomass is low.

Leslie *et al.* (1983) detected a significant increase in nitrate-nitrite concentrations in two Florida lakes following macrophyte removal by grass carp. Increased alkalinity may have been related to decreased photosynthetic activity and the resulting release of carbonate ions following the elimination of macrophytes. Mitzner (1978) reported a significant increase in alkalinity following a 91% reduction of aquatic macrophytes by grass carp in a lake in Iowa. Elimination of macrophytes in a cooling reservoir in North Carolina using *Tilapia zillii* also resulted in an increase in nitrate/nitrite and alkalinity concentrations (Crutchfield *et al.*, 1992). An intensive decomposition of macrophytes of their residues may result in an increase in ammonia, the concentration of which, if exceeding 1.0 mg L^{-1}, would stop grass carp to feed (Kracko and Noble, 1993).

In experiments in aquaria the macrophyte *Anacharis canadensis* was eliminated by the grazing activity of rudd (Hansson *et al.*, 1987). The grazing resulted in an increase in phosphorus concentration in water. In ponds this caused an increase in the biomass of epiphytic algae but no increase in phytoplankton.

Common carp reworking and digestion of bottom sediments may lead to a substantial increase in phosphorus in water. Lamarra (1975) suggested that a carp population of 200 kg ha^{-1} could internally load a lake with orthophosphate at 0.52 mg P m^{-2} day^{-1}. This might result in an increase in phytoplankton where plant control using common carp is proposed.

Aquatic Macrophytes versus Phytoplankton

In southwestern Finland, the small and shallow Lake Littoistenjarvi alternates between two alternative states: mesotrophic, characterised by well-developed stands of aquatic macrophytes, and eutrophic, when the macrophytes decline. The high abundance of submersed macrophytes and clear water in the early 1990s, was followed in 1992 by high nutrient and chlorophyll levels, high turbidity and reduced stands of macrophytes. The situation changed again during 1993–1996 when the lake water was clear, chlorophyll levels low relative to total phosphorus, and aquatic macrophytes moderately abundant. The roach stock was moderately strong (71 kg ha^{-1}) after the high abundance of submersed macrophytes in the early 1990s, declining to about 28 kg ha^{-1} by 1996. Perch biomass was notable, but consisted mainly of large (>20 cm) individuals that were probably piscivorous (Sarvala *et al.*, 1998). In 1994–1996 the total fish biomass did not decrease, but there was a clear shift in the size distribution towards larger fish; especially small roach were almost absent in 1995–1996.

This size shift probably led to a lowered predation pressure on zooplankton, and the cladoceran biomass showed some increase in 1994–1996. In this Finnish lake cladoceran zooplankton was thus controlled by fish predation, but water quality was not dependent on zooplankton. Phytoplankton abundance was more tightly linked to long-term fluctuations of submersed plants. The abundance of large crustaceans in 1993 (and an almost absence of *Daphnia*) and in later years of the littoral cladocerans *Sida* and *Simocephalus*, probably originated from the submersed vegetation, which may also have offered a refuge from fish predation for other fish. Interestingly, during 1993–1996, phytoplankton biomass remained low in spite of very low crustacean zooplankton biomass during most of the summer. Similar role of submersed aquatic plants has been also reported for some other lake ecosystems (Blindow *et al.*, 1993; Schriver *et al.*, 1995).

Periphyton Function

Sand-Jensen and Borum (1984) and Sand-Jensen and Sondergaard (1981) belong to a number of workers who pointed out the importance of epiphytic (synonym: periphytic) algae coating submersed macrophytes as a factor which leads to the suppression of aquatic plants. Excessive growth of epiphytes is harmful to the macrophyte, because insufficient quantities of light and inorganic carbon (HCO_3 and CO_2) can reach the plant and consequently the macrophyte photosynthesis is hampered (Sand-Jensen and Sondergaard, 1981). Epiphytic algae are more efficient at obtaining nutrients and light due to their large surface area to volume ratio, high biotic potential and closer proximity to the water column. They may also have more efficient nutrient uptake mechanism (Maberly and Spence, 1983). Periphyton provides more nutritious substrates than sediments or vascular plant tissues, and consequently it is an important energy source for both detritus and grazing food chains (Gressens, 1995).

Epiphytic algae benefit from inorganic nutrients and dissolved organic compounds leaking from the host macrophytes, and these may be readily used and beneficial to algal and bacterial communities of the epiphyton, especially in oligotrophic waters (Bronmark and Vermaat, 1998). Difference in algal communities on different species of macrophytes under low nutrient availability has been suggested to be due to host-specific composition of the released organic and inorganic compounds (Eminson and Moss, 1980).

Periphyton is grazed by a diversity of both invertebrate and vertebrate aquatic organisms. The invertebrates include molluscs, nematodes, microcrustaceans (Cladocera, Copepoda, Ostracoda), amphipod and isopod crustaceans, crayfish, mysids, chironomid larvae, caddisfly and mayfly larvae. Invertebrate grazers may play an extremely important role in controlling periphyton. Grazing snails benefit macrophytes by removing the epiphytic cover, as shown in experiments by Bronmark (1985): the growth rate of *Ceratophyllum demersum* increased in the presence of snails by almost 30 % due to the removal of epiphytic cover. Similar results were obtained by Martin *et al.* (1992) in Bays Mountain Lake (Tennessee, USA) where the biomass of submersed macrophytes (*Najas flexilis* and *Potamogeton diversifolius*) increased 60x in cages where snail-feeding redear sunfish and small sunfish (*Lepomis macrochirus* and *L. microlophus*) were excluded.

The results suggested a chain of strong interactions (i.e. from fish to snails to periphyton to macrophytes) that may be important in lake littoral system. This contrasts sharply with earlier predictions based on cascading trophic interactions that propose that fish predation on snails would enhance macrophyte biomass.

In an experimental study on a small pond in southern Sweden, with dense submersed vegetation dominated by *Elodea canadensis*, with much less abundant *Potamogeton pectinatus, P. obtusifolius, P. crispus, Ranunculus trichophyllus*, and *Chara* sp., the presence of tench indirectly increased the biomass of periphyton through a reduction of grazing pressure by snails, on which the fish fed (Bronmark, 1994). A similar study carried out on two North American lakes showed that the highly specialized moluscivorous fish *Lepomis gibbosus* dramatically reduced the density and biomass of snails, which in turn resulted in an increased biomass of epiphyton (Bronmark *et al.*, 1992). These studies showed strong cascading interactions between fish, snails, algae and submersed macrophytes. Given the difference in shade tolerance between plant species, cascading trophic interactions initiated by a predatory fish may well influence distribution patterns and species composition of submersed macrophyte communities in lake littoral zones. With an increasing density of predators feeding on snails the most shade-intolerant macrophytes are predicted to disappear first, and at high levels of predation, possibly coupled with increased nutrient enrichment, the entire submersed macrophyte community may disappear.

Intermediate-sized herbivores - fish and crayfish - reduce macrophytes and periphyton similarly, with the literature suggesting that the periphyton consumed by these species is primarily filamentous macroalgae that may be intentionally ingested by the herbivores. But Scott and Crossman (1973) pointed out that many American fish are algivorous or use filamentous algae as a substantial food source, particularly in late summer. In the slow-flowing rivers of southeast

England young-of-the-year roach (*Rutilus rutilus*) and chubb (*Leuciscus cephalus*) switch to feeding on periphyton when preferred zooplankton is not available, either due to low numbers or at night when visual feeding is precluded. This switch coincides with a decline in the growth rate of these fish (Jones *et al.*, 1998). By contrast, the smallest herbivores-snails and insects - have little (insects) or no impact (snails) on macrophytes but do substantially reduce periphyton, especially microalgae (Lodge *et al.*, 1998).

The seasonal timing of the presence of different grazers may differ substantially. Chironomid larvae often dominate in spring, and graze until their pupation and emergence. Snails are present all year around and may remain active even at low temperatures. In some water bodies oligochaetes appear to be the most important herbivores in terms of overall grazing impact (Bronmark and Vermaat, 1998). But seasonality may be weakly expressed, or completely lacking in tropical waters. Thus, the consequences of grazing may differ in subtropical and tropical lakes from those in temperate water bodies. Botts and Cowell (1993) found that abundance of epiphytic algae and invertebrate grazers were only weakly correlated in a subtropical lake, whereas in temperate lakes they usually show strong temporal correlation.

In Argentina, the three rivers, Uruguay, Paraguay and Parana, form Rio de la Plata. The characid fish *Prochladius lineatus* is the dominant species among the some 300 species that inhabit Rio de la Plata. It represents around 60% of the biomass and over 70% of fish catches. It has been suggested that one of the causes of the extraordinary success of this detritivorous fish is its capacity to selectively feed on nutritionally superior detritus, i.e. epiphytic detritus, rich in total amino acids. Also, it has been suggested that it might obtain periphyton associated with roots and shoots of aquatic macrophytes. As water hyacinth covers 80% of lakes and ponds of the middle Parana floodplain, Planas and Neiff (1998) investigated this plant, collected from the Parana River floodplain, for periphyton. They found that mats of water hyacinth let through insufficient light to allow full development of periphyton.

Weisner *et al.* (1997) investigated a combined impact of epiphytes and waterfowl grazing on submersed vegetation on three Swedish eutrophic shallow lakes in the clear-water state. They found that mechanisms hampering submersed vegetation were strongest at shallow and/or sheltered locations. The growth of *Myriophyllum spicatum*, planted in the same substrate and at the same water depth, was compared between sheltered and wave exposed sites in two lakes. After 6 weeks the plants were significantly smaller at the sheltered sites, where periphyton production was about 5 times higher than at the exposed sites. Enclosure experiments were conducted to evaluate the effects of waterfowl (mainly coot *Fulica atra* and mute swan *Cygnus olor*) grazing on macrophyte biomass. *Potamogeton pectinatus* growth was decreased by grazing, whereas *M. spicatum* was not affected. The effects were greater at a sheltered than at a wave-exposed site. The results suggest that competition from epiphytes and waterfowl grazing hamper the development of submersed vegetation at sheltered and/or shallow locations.

Zooplankton

Aquatic macrophytes provide shelter or refuge to zooplankton, and the presence of especially large zooplankton controls small phytoplankton. If the larger zooplankton becomes scarce, the phytoplankton community will change to slow-growing bigger, mainly inedible blue-green species. These bigger, often colonial algae compete more successfully with the small algae for nutrients and establish high-density populations, sometimes leading to algal blooms. In turn, this restricts light penetration into deeper water and leads to the death of submersed macrophytes.

In the English lake Little Mere experiments were conducted to investigate the relative importance of top-down (zooplankton grazing) and bottom-up (nitrogen-limitation) control in limiting algal growth, and the role of aquatic macrophytes in this process (Stephen *et al.*, 1998). It is generally recognized that submersed macrophytes provide a refuge for cladocerans and a habitat for plant-associated macroinvertebrates, both of which graze on phytoplankton and epiphytes. However, in experiments with a high stickleback (*Gasterosteus aculeatus*) density the fish controlled the zooplankton associated with the macrophytes. The same conclusion was reached by Jeppesen *et al.* (1998), who found that the refuge effect of submersed macrophytes almost disappeared - even for small-sized cladocerans - if fish density exceeded ca. 4 fish m^{-2}. However, if plant density is high and fish are not forced into the vegetation, pelagic zooplankton may aggregate in the submersed macrophytes during the daytime. Floating-leaved plants and reed belts have a comparatively low refuge effect due to low stem density.

The efficiency of plant beds as refuge for zooplankton is often high in eutrophic lakes. In less eutrophic and mesotrophic lakes, the impact of predatory fish increases. In Europe, in such lakes perch predation becomes more important, and the foraging conditions for predatory fish improve, due to, for instance, increased transparency. Young planktivorous prey fish thus seek refuge in the vegetation, which does not provide effective refuge from the predator (Jeppesen *et al.*, 1998). In mesotrophic-oligotrophic lakes the predatory pressure on planktivores becomes even stronger.

In some other English lakes (Cromes Broad, Hoveton Little Broad/ Pound End, and Upton Broad) *Daphnia* spp exhibited an early summer peak of abundance but had declined rapidly by July, probably as a result of predation by fish. In dense macrophyte stands *Daphnia* spp persisted after their elimination in open water, indicating some refuge effect (Stansfield *et al.*, 1997). At sites with macrophyte cover and/or low fish predation pressure, *Daphnia* was replaced by *Ceriodaphnia*, both with or without *Simocephalus*, thus maintaining large populations of grazing Cladocera, apparently exerting a high grazing pressure on phytoplankton. This occurred even under high predation pressure by fish. The authors have suggested that dense macrophytes offer the most suitable refuge for cladocerans through provision of predator-free space.

Cascading trophic interactions are known to play an important role in the transfer of fish predation effects via zooplankton to the phytoplankton community, and this may affect also the structure and function of the microbial community. Jurgens and Jeppesen (1998) followed the cascading effects on microbial food web structure in a dense macrophyte bed (mainly *Potamogeton pectinatus*) under experimental conditions. As a result of differing fish predation pressure (strong where no macrophytes present, low where macrophytes present), there was a high density of different cladocerans and cyclopoid copepods in the enclosure system with macrophytes, whereas ciliates, rotifers, and cyclopoid copepods dominated in the enclosures without macrophytes. In the macrophytes the abundance of phytoplankton, protozoans and bacteria was very low, whereas with no macrophytes a diverse assemblage of pico-, nano- and microplankton coexisted with the metazooplankton community.

Ciliate density was approximately two orders of magnitude higher in the enclosures without macrophytes. The protozoans there probably exerted considerable grazing pressure on the bacterial and algal communities. Bacterial concentrations were four- to fivefold higher in enclosures without macrophytes than those with macrophytes. The experiments have shown that fish predation impact can cascade from fish to the bacterial level, and that this is probably a general feature of more eutrophic systems, such as shallow eutrophic lakes in the temperate climate of the northern hemisphere.

Aquatic Macroinvertebrates

Aquatic macrophytes are an important habitat for aquatic macroinvertebrates. The plants provide protection from predators and current, and are a direct and indirect source of food of fish and aquatic birds. In plant beds, benthic invertebrates are generally more numerous and more diverse than in open water. Diehl and Kornijow (1998) showed the increase in epiphytic macroinvertebrates with increasing macrophyte densities, and how this is paralleled by an increase in fish biomass. Species composition and biomass of macroinvertebrates usually differ from plant species to species, depending on the structural complexity of the plant, changes in the morphology of macrophytes during the growing season, amount of the periphyton and detritus on plants, changes in water level, life cycle of the individual macroinvertebrate species, and some other factors. Differences in invertebrate richness, biomass and distribution will have a direct effect on the fish populations depending on aquatic macroinvertebrates as source of food. Indirectly, macrophytes are also important for the benthic macroinvertebrates, for which they provide protection against foraging by fish and aquatic birds, as well as providing the invertebrates with a supply of organic matter from the decomposing plants.

Kornijow and Gulati (1992) investigated aquatic macroinvertebrates associated with aquatic vegetation about 18 months after the biomanipulation of Zwemlust, a small Dutch lake. They found that the macrofauna did not differ notably from that in non-manipulated lakes. The species diversity differed on the different species of aquatic plants. The greatest diversity was on *Chara* sp., moderate on other two submersed plants *Elodea nuttallii* and *Ceratophyllum demersum*, and on *Phragmites australis*, and the lowest on filamentous green algae, primarily *Mougeotia* sp. Some taxa were restricted to certain plants, e.g. the mollusc *Acroloxus lacustris* found solely on reed, while the molluscs *Armiger crista, Planorbis corneus, Valvata piscinalis* and leeches, except *Erpobdella* sp., were recorded only on the submersed plants. Larvae of Tipulidae and Ephydridae occurred exclusively on filamentous algae.

The density and biomass of the fauna per 100 g of plant wet weight on different plant species markedly differed, with the lowest values on reed and among *Mougeotia.* Much higher density and biomass occurred on submersed plants, especially on *Chara* sp. Generally, the densities of the phytomacrofauna were relatively low but the biomass values high, compared with many other lakes. Gastropoda, larvae of Chironomidae and Trichoptera were the main quantitative components of the phytomacrofauna. It was noted that the greater the fragmentation of macrophyte leaves the greater the faunal density, and usually higher the taxonomical variability of the associated animal populations. In this respect, Lake Zwemlust aquatic vegetation confirmed findings from some other lakes. There was a high degree of similarity between the fauna associated with *Ceratophyllum* and algal mats, with a high dominance of the chironomid *Cricotopus* larvae. The larvae feed primarily on the algae constituting mats and those covering the macrophyte (Kornijow and Gulati, 1992).

During strong stratification, aquatic macrophytes provide a diversity of microhabitats and refugia for aquatic macroinvertebrates which are temporarily or permanently excluded from the lake bottom by low dissolved oxygen concentrations. Differences in the habitats provided by the floating *Pistia stratiotes* and submersed *Ceratophyllum demersum* in Volta Lake (Ghana, West Africa) were reflected in species composition, dominance and standing crops of their associated aquatic macroinvertebrates. While grazers and filter-feeders, such as zygopteran nymphs and chironomid larvae, dominated *Ceratophyllum*, submersed *Pistia* roots were utilised by predatory anisopterans, the hemipteran *Diplonychus*, and the grazing molluscs *Bulinus forskali* and *B. truncatus rohlfsi* (Petr, 1968).

The density of macroinvertebrates in *Pistia* roots reached a maximum of about 16,000 individuals m^{-2}, which is close to the density of 18 620 individuals m^{-2} collected by Pelli and Barbosa (1998) from *Salvinia molesta* roots in a lake in Brazil. The dry mass of macroorganisms on *Pistia* at 12 g m^{-2} was higher than that reported in the literature for floating plant associated invertebrates from African, Asian or South American lakes (McLachlan, 1969; Lim and Furtado, 1975; Junk, 1973). The association of *Ceratophyllum* with shallows was found to represent a protective mechanism for the associated macroinvertebrate fauna: wind-blown *Ceratophyllum*, carried off-shore and found stranded in the crowns of drowned trees several hundred metres from the shore, supported considerably lower animal densities than those in the shallow littoral (Petr, 1968).

The free-floating submersed *Ceratophyllum demersum* is a common plant in tropical African waters. While it can be a nuisance plant as it provides a habitat for host snails of the parasite *Schistosoma haematobium*, it is considered to be beneficial for fish as it serves as a substrate for aquatic invertebrates and periphyton, on which many fish feed. Due to the vertical distribution of this plant, which may reach several metres of length, Petr (1968), rather than making an attempt of expressing the number and biomass per square meter, which would be irrelevant because a different length/depth of the individual plants, expressed the abundance and biomass of invertebrates associated with *Ceratophyllum* as per 1 000 g of dried plant.

During three months of observations, the inshore *Ceratophyllum* contained 16 502, 39 020, and 92 156 organisms, respectively (corresponding to 34, 40 and 76 g wet weight), as compared to 12 095, 6 891, and 6 234 organisms (corresponding to 7.6, 5.9 and 5.2 g wet weight) in off-shore *Ceratophyllum* plants, stranded on drowned trees. The off-shore, usually isolated individual plants, are exposed to heavy fish predation, and their recolonisation appears to be minimal. If using conservatively the biomass per 1 000 g of dried plant as a biomass per m^2 for the inshore habitat, the biomass exceeds twice the maximum biomass of invertebrates found on *Pistia* roots in the same lake, i.e. 35 g m^{-2}. The rich aquatic invertebrates of aquatic plants in Volta Lake must play an important role in the high fish production in this reservoir.

High densities of macroinvertebrates associated with aquatic macrophytes have been recorded from the borders of macrophyte beds on floodplains of Orinoco (Venezuela) and Parana (Argentina), and there was considerable difference between the 'isolation' phase, i.e. when the floodplains were not connected with the river, flooding phase, and the inundation period (Blanco-Belmonte *et al.*, 1998). In Orinoco floodplain lakes, where macrophytes develop only during the rainy season, there is only one density peak. The peak densities there were: for water hyacinth 38 000 org. m^{-2}, 68 000 org. m^{-2} for *Paspalum repens*, and up to 600 000 org. m^{-2} for *Nymphaea*. Such high densities represent lucrative feeding grounds for numerous fish.

In a review on herbivory on aquatic macrophytes Lodge (1991) pointed out that while a majority of herbivorous invertebrates eat periphyton, many do eat macrophytes. Those who eat macrophytes may be functionally important in food webs and nutrient fluxes. As a trophic link they are significant in community and ecosystem functions.

Some caddisfly larvae are known to be among the few aquatic invertebrates directly feeding on aquatic macrophytes. High densities of herbivorous insects, such as a caddisfly, several moths, several weevils, and one or more chironomid species, are known to accompany the decline in *Myriophyllum* (Smith, 1991). *Myriophyllum spicatum* and *Hydrilla verticillata* have become serious aquatic weeds in many parts of continental USA, and a search is going on for effective insect biological control agents. Balciunas (1991) and Balciunas *et al.* (1989) reported on their search for such agents in China and Australia, which followed Balciunas previous search in Africa.

In Lake Tanganyika, *Polypedilum* chironomid larvae damage apical meristem of *Hydrilla*, and together with grazing by fish, are the two factors mainly responsible for containing *Hydrilla* (Markham, 1986). Whether the emphemeropteran nymphs of *Povilla adusta*, which burrow in *Polygonum* stems in Volta Lake (Petr, 1970) and in rootlets of *Pistia stratiotes* in Lake Chad (Dejoux, 1969) cause a serious damage to their host plants, is not known. They do not feed on them and use them as shelter from which these larvae filter phytoplankton from the surrounding water, or graze on the periphyton. The nymphs are known to burrow into bark and wood of trees, which speeds up the destruction of trees inundated in some African tropical reservoirs.

In Lake Victoria and Lake George (both in Uganda), where the nymphs burrow into papyrus plants, they also attack wooded canoes of fishermen, or even boats made of fibreglass. The nymphs require for their existence a relatively high dissolved oxygen concentration in the surrounding water, which limits their lateral distribution within the fringing papyrus swamp to only about 1.5 m distance from the edge.

In laboratory conditions the trichopteran *Anabolia nervosa* larvae were found to show a preference for young leaves of *Potamogeton perfoliatus*, which were richer in nitrogen than leaves of terrestrial trees *Alnus glutinosa* and *Fagus sylvatica* (Jacobsen and Sand-Jensen, 1994). The larvae consumed 3 times more fresh *Potamogeton* than *Alnus* leaves, but the same amount in terms of dry mass. Larvae grown on *Alnus* accumulated more fat than larvae grown on *Potamogeton*, while the reverse was true for protein accumulation. The study demonstrated that fresh tissue of *Potamogeton perfoliatus* provided a valuable source to this caddisfly larva. In the field, i.e. a Danish stream, however, the plant constituted only 1–5% of the food intake by this trichopteran (Jacobsen and Sand-Jensen, 1994a). The highest plant consumption was from early May to mid-June, but the invertebrate herbivores harvested only 1.3 to 1.8% of the total annual plant production. Otto and Svensson (1981) discovered in experiments that the caddisfly larvae of *Sericostoma personatum* (Spence), found in Swedish waters, consumed significantly less leaves with high nitrogen content than leaves with low nitrogen content, and suggested that the low palatability of aquatic macrophytes may be due to chemical defense mechanisms. Larvae of the trichopteran *Potamophylax cingulatus* were also tested and confirmed that aquatic plants produce secondary plant substances, which reduce the attacks by aquatic herbivores.

A study of invertebrates associated with three macrophytes in the Macquarie River, Tasmania (Australia) showed major differences in the invertebrate assemblage in any one month, and also a large turnover of taxa associated with the species of macrophyte species from one month to the next (Humphries, 1996). The three macrophytes sampled were *Myriophyllum simulans/variifolium*, *Triglochin procera* and *Eleocharis sphacelata*, each one differing in the structural complexity and its location in the littoral zone. The greatest abundance of macroinvertebrates was found associated in all months (and at all water levels) in the structurally complex and shallowest *Myriophyllum*. However, this plant had a much lower number of taxa than the structurally simpler and deeper water *Triglochin* and *Eleocharis*. The invertebrate abundance and *Myriophyllum* biomass were found to initially increase together, but as the plant biomass became very high, i.e. the plant became dense, the invertebrate abundance declined. This may have an implication for fish feeding, which in such a case is limited by the plant density and the availability of the prey. A similar situation was also found on *Myriophyllum*-associated mosquito *Anopheles*, with the density of the mosquitoes decreasing with an increase in the stem density (Orr and Resh, 1992).

Cebrian and Duarte (1994) found that consumption of freshwater macrophytes, seegrasses and terrestrial grasses by invertebrates and vertebrates ranged fom 1 to 80% of plant production. In two studies of floating-leaved freshwater macrophytes the invertebrate consumption of plant production was less than 10% for *Nymphoides peltata* (Van der Velde *et al.*, 1982), and 3–6% for *Nuphar lutea* (Setala and Makela, 1991). A number of authors as quoted in Jacobsen and Sand-Jensen (1994) have observed a heavy damage of submersed and floating-leaved macrophytes by invertebrate herbivores.

The presence of vegetation serves as a food resource and cover for many vulnerable macroinvertebrates in the mud, which otherwise would have been quickly eliminated in the absence of plants (Petridis, 1990). McLachlan (1969) found an increase in the number of pulmonates and bivalves in the mud of Lake Kariba, a reservoir in eastern tropical Africa, following an invasion of aquatic macrophytes. But Killgore *et al.* (1989) found that dense monospecific beds of *Myriophyllum spicatum* contained few benthos and could support less diverse fish community. In the USA, on the other hand, in ponds without vegetation, common carp and bluegill greatly reduced the abundance of oligochaetes and nematodes (Forester and Lawrence, 1978).

Availability of aquatic invertebrates as fish food depends on the effectiveness of the shelter provided to them by aquatic macrophytes, and their productivity. Their productivity will depend on the food availability. Epiphyton, the source of food of the majority of macrophyte dwelling macroinvertebrates, is at its optimal development when the surrounding water has intermediate nutrient concentrations (Carpenter *et al.*, 1998).

The most important aspects of the periphyton-invertebrate grazer-fish interaction in lakes is that herbivorous invertebrates provide an alternative food source for fish and hence reduce the pressure on zooplankton (Jones *et al.*, 1998). Without such invertebrates, the ontogenetic shifts seen in feeding adult fish would not occur and all fish would compete for the same food resource, a situation seen in plantless eutrophic lakes. Such feeding on invertebrates by fish will result to some extent in an increase in periphyton, but the overall cost to the plants may be outweighed by gains resulting from increased zooplankton abundance (Jones *et al.*, 1998).

Insects, which are used for biological control of nuisance aquatic plants, may also become food for fish, thus negating their impact. To control *Myriophyllum spicatum*, the aquatic weevil *Euhrychiopsis lecontei* was released in Lake Auburn (Minnesota, USA). Fish stomach analysis of the sunfish (*Lepomis macrochirus*) and pumpkinseed (*Lepomis gibbosus*) has shown a frequency occurrence for the weevil larvae and adults of 10.3% to 28.6% (Sutter and Newman, 1997). The findings suggest that sunfish and pumpkinseed predation may decrease the effectiveness of the biological control of *Myriophyllum* using *E. lecontei.*

Additional information on herbivore and detritivory on freshwater macrophytes by invertebrates is given in a review by Newman (1991).

Crayfish

The proportion of aquatic macrophyte in crayfish food may represent up to 75% of their total food intake as documented by a number of workers since the 1980s. Crayfish can also substantially reduce or eliminate submersed vegetation, as documented, for example, from lakes and ponds in the northern USA and from Lake Naivasha, Kenya. In New

Mexico the crayfish *Orconectes causeyi* successfully cleared several lakes of *Potamogeton*, *Elodea*, *Ranunculus* and *Myriophyllum*, but avoided *Polygonum*, *Scripus* and *Typha* (Schuytema, 1977, and references therein). In Arizona's Parker Lake, when the crayfish (*Orconectes causeyi*) population increased, their intensive grazing on submersed plants reduced the protective cover, which in turn increased the crayfish vulnerability to the predatory largemouth bass. As the predation by largemouth bass increased, the population of crayfish decreased, which led to the recovery of *Myriophyllum* and an increase of crayfish (Saiki and Tash, 1979). Investigations of prey capture rates of trout (*Oncorhynchus mykiss*) feeding on two species of amphipods in an environment also rich in crayfish (*Orconectes propinquus*) and smallmouth bass (*Micropterus dolomieu*) have shown that prey capture rates of trout fell drastically with increasing cover. Capture rates of crayfish by smallmouth bass also declined (Crowder and Cooper, 1979a).

In England, the introduced signal crayfish *Pacifastacus leniusculus* fed on the following five main diets, in order of their importance: vascular detritus, filamentous green alga *Cladophora*, crayfish fragments, chironomid larvae, and mayfly nymphs (Guan and Wiles, 1998). The larger the crayfish, the more vascular detritus was ingested. The frequency of occurrence of *Cladophora* was increased with crayfish size in the winter.

Reductions of macrophyte root or shoot biomass by grazers range up to 100%, with reductions exceeding 50% reported for crayfish, insect larvae, snails, fish, and waterfowl (Lodge, 1991). In laboratory experiments, non-consumptive destruction by *Orconectes rusticus* comprised 30–95% of cut macrophyte biomass, with the percentage differing among macrophyte species. *O. rusticus* is known to feed selectively, as tested on 14 different plants species by Lodge (1991). Evergreen aquatic plants are not preferred, while those preferred are usually perennial. Avoidance of evergreens by grazers is also a common pattern among terrestrial grazers (MacLean and Jensen, 1985). This appears to result from secondary chemical defenses.

Chambers *et al.* (1990, 1991) studied the impact of crayfish introduction on submersed plant communities and associated aquatic macroinvertebrates. In feeding preference trials with *O. virilis*, out of ten submersed aquatic plant species *Chara* sp. and *Lemna trisulca* were the preferred food, followed by *Myriophyllum exalbescens* or *Utricularia vulgaris*. Chemical analysis of the plants have suggested that the feeding preference of this crayfish is given by the ease of handling small, short, bottom-dwelling plants (*Chara* and *Lemna*), and not by the plant chemistry. The same species of crayfish had a significant impact on both the aquatic macrophytes and macroinvertebrate community, even at a low biomass of less than 20 g m^{-2}.

Tests carried out under semi-natural conditions in plastic pools containing *Potamogeton richardsonii*, *Myriophyllum exalbescens*, *Nuphar variegatum* and *Sparganium eurycarpum* have shown that there was a major difference between the impact of the females and the males on the vegetation. While the authors described the effect of female crayfish as generally stimulatory, leading to an increase in the biomass of *Myriophyllum* and *Potamogeton*, and to an increase in *Myriophyllum* length, the plant growth was negatively impacted by male crayfish. The experiments have also shown the feeding selectivity of the crayfish: of the four macrophytes only *Myriophyllum* was consumed, while the crayfish just clipped the shoots of the other three plants which were left to float on the water surface. Female crayfish *O. virilis* were found to virtually eliminate snails from the surfaces of the macrophytes. Thus, the observed increase in the biomass of *Potamogeton* and *Myriophyllum* seemed to be related, at least in part, to the reduction in herbivorous snail abundance. Crayfish thus seems to indirectly alter the macroinvertebrates by changing the abundance and species composition of the aquatic macrophyte community.

Cronin (1998) followed the feeding choices of the Louisiana crayfish *Procambarus clarkii*, a commercially valuable species now cultivated in a number of other countries. When plants were offered as fresh tissue, *Procambarus* consumed large amounts of *Chara*, *Ceratophyllum* and *Spirogyra*; intermediate amounts of *Typha*; and little (if any) *Potamogeton*, *Batrachospermum*, *Nuphar*, *Carex*, or *Iris*. The three preferred species are finely branched or filamentous suggesting that plant structural properties are important determinant of *Procambarus* feeding decision. Cronin (1998) also tested crude extracts from plants: those from *Ceratophyllum*, *Nuphar* and *Iris* significantly reduced consumption, while extract from *Typha* significantly stimulated feeding. But according to Bolser *et al.*

(1998), while lipophylic crude extract from *Typha* significantly deterred feeding of this species of crayfish, when offered it in an agar-based diet lacking structural defenses, the crayfish showed great preference for it. According to Cronin (1998) among plants without chemical defense mechanisms *Procambarus* will target those with high nitrogen content. Cronin concluded that generalist herbivores such as *Procambarus*, base their feeding decisions on multiple plant traits such as morphology, structure, chemical defenses, and nutritional value. *Procambarus* will eat only plants that they can handle, shred, and digest.

In experiments (Nystrom and Perez, 1998), crayfish *Pacifastacus leniusculus* had a strong impact on snail mortality. Their study suggested that crayfish can structure the abundance and size distribution of thin-shelled snails (*Lymnaea stagnalis*) through size-selective predation and reduction of macrophytes. Hanson *et al.* (1990) observed a decrease in the density of amphipods and the total biomass of non-molluscan invertebrates at male crayfish biomasses of 5 and 10 g m^{-2} and suggested that this may be directly related to crayfish predation or indirectly to reduction in macrophyte biomass and species richness. In laboratory experiments with *O. rusticus* and *O. virilis* Olsen *et al.* (1991) observed that *O. rusticus* damaged more snails than *O. virilis*, and *O. rusticus* had a greater weight-specific ingestion rate than *O. virilis* on snails but not macrophytes.

Feeding beef liver to *O. rusticus* significantly reduced consumption and non-consumptive destruction of macrophyte *Vallisneria* (Lorman, 1980). Lodge and Lorman (1987) suggested that the presence of snails probably reduces macrophyte consumption by *O. rusticus* in the littoral zone of northern Wisconsin lakes. In Australia in experiments carried out by Whisson (1997), *Vallisneria* sp. was found to have a protective function for juvenile and adult crayfish marron (*Cherax tenuimanus*), reducing the predatory impact on them of silver perch (*Bidyanus bidyanus*). The author suggested that there could be a useful role for this plant in commercial marron production.

Changes in macrophyte communities due to both the direct and indirect effects of crayfish grazing may, in turn, affect fish communities, through the loss of the opportunity of some fish to hide from predators, change in or shortage of suitable spawning grounds, changes in or loss of food organisms such as periphyton, macroinvertebrates, zooplankton. In Wisconsin lakes, the introduction of *O. virilis* has been linked to detrimental changes in the fish community (Lodge *et al.*, 1985). Carpenter and Lodge (1986) summarized the impacts of omnivorous crayfish on a lake ecosystem. Rickett (1974) warned of the possibility of excessive vegetation elimination in lakes or ponds by uncontrolled crayfish populations. In northern Wisconsin lakes, the probably inadvertent introduction of the crayfish *Orconectes rusticus* and the subsequent increase in its numbers went in parallel with a decline in submersed macrophytes (Lorman, 1975). In Lake Metonga, the disappearance of macrophytes stopped the recruitment of the fish walleye (*Stizostedion vitreum*) (Lodge *et al.*, 1985), and in some other northern Wisconsin lakes the indigenous crayfish *Orconectes virilis* and *O. propinquus* were displaced by the intruder (Olsen *et al.*, 1991). Where the population of crayfish crashed, as in Lake Naivasha, Kenya, due most probably to overfishing of the introduced *Procambarus clarkii*, this was followed by the recovery of rooted aquatics (Harper *et al.*, 1990).

Crowder *et al.* (1998) constructed a simplified generalized food web for submersed macrophyte system. Boxes contain the major types of interactors. Lines ending in arrows indicate positive effects to the upper trophic level of energy flow up the web. Lines ending in closed circles indicate negative effects on the lower trophic level due to the level above. These may include the potential for predators to "control" prey densities. Dashed lines indicate potential interaction such as leaking of nutrients from macrophytes to epiphytes or indirect effects of cover on predator-prey interactions. Dotted lines indicate possible routes of nutrient recycling from consumers. The authors point out that individual systems will diverge from this diagramme, and many species shift from one role to another through ontogeny. In many systems, the herbivores include both mesograzers (primarily oriented toward epiphytic algae) and macrograzers (primarily oriented toward macrophytes). In systems dominated by macrograzers, the hypothetical effects of reduction of large predators would be an increase in macrograzers, which could lead directly to losses of macrophytes along with their associated epiphytes and fauna.

PISCIVOROUS FISH - PREY RELATIONSHIPS AND AQUATIC MACROPHYTES

A study by Chick and McIvor (1994) on the abundance and fish species composition reconfirmed that beds of different macrophyte types comprise distinct microhabitats for juvenile and forage fish. From the perspective of these fish, the littoral zone of Lake Okeechobee, the largest lake in Florida, may be better described as a landscape of microhabitats with beds of different macrophyte types forming microhabitat patches comprising a mosaic. The fish abundance was related to macrophyte type, but not to vegetation standing crop or water depth. They suggested a number of mechanisms to be responsible for the fish distribution.

For example, juvenile and forage fish feed on a variety of invertebrates, and the distribution of invertebrates are known to differ with macrophyte type (e.g. see Hargeby, 1990). Differences in plant surface area and vegetation decomposition could impact epiphytic communities and detrital food webs associated with these macrophytes (Paterson, 1993). Predation pressure probably also differs among macrophyte types because differences in vegetation density can greatly affect the ability of predators to capture prey (Savino and Stein, 1992). Macrophyte structural complexity can affect both the ability of predators to capture small fish and the ability of small fish to capture invertebrate prey. Vegetated habitats tend to be populated primarily by small forage fish and juveniles of larger species, which use these habitats as a refuge from predators and feeding grounds. Chick and McIvor (1994) suggested that a landscape view of littoral zones will prove useful for applied lake management, such as habitat manipulation and littoral zone restructuring, as well as for ecological investigations on fish population dynamics and lake processes.

In the littoral zone, aquatic macrophytes are thought to enhance the survival probability of prey fish by decreasing encounter rate or predation rate (Hosn and Downing, 1994). Moderately complex aquatic plants are important for young fish to hide and feed. Such plants will produce more stable prey-predator relations between fry and piscivorous fish (Savino and Stein, 1982). In Dutch water bodies, survival of young *Esox lucius* was higher in vegetated areas than in areas without aquatic macrophytes (Grimm, 1983). In a Florida lake in the presence of a coverage of 75% of *Hydrilla verticillata* the survival of the young *Esox niger* and some other American sport fish was higher than outside such areas (Shireman *et al.*, 1983). Small bluegill (*Lepomis macrochirus*) have been shown to be restricted to dense vegetation in the presence of piscivorous fish (Werner *et al.*, 1983). Submersed aquatic macrophytes provide cover for young largemouth bass, and bass (*Micropterus salmoides* and *M. punctulatus*) grow better in intermediate to high plant densities of *Hydrilla* and edge areas than in areas of uninterrupted high-density plants or areas without plants (Morrow *et al.*, 1991).

This is consistent with other studies that have shown aquatic plants to be important to the growth of fish. However, several studies have shown that the growth rate of prey fish is retarded when they are confined to the vegetation due to predation risk (Mittlebach, 1988; Persson, 1993; Persson and Eklov, 1995). The maximum foraging rate decreases when the fish shifts from the pelagic to the vegetation habitat (Persson and Crowder, 1998). The slower growth results from an increase in competition among the prey fish in the refuge.

The presence of vegetation will also affect the foraging mode of specific piscivorous predators. For example, perch and largemouth bass change from an active pursuit foraging mode to an ambush sit-and-wait foraging mode with an increase in vegetation density (Savino and Stein, 1982, Eklov and Diehl, 1994).

In many lakes throughout the north-central United States, the centrarchid bluegill sunfish make up most of total fish biomass (Mittelbach and Ossenberg, 1993). This species hatches in the littoral and moves to open water for a few weeks before moving back to the sheltered vegetation habitat. As an adult, it feeds on zooplankton in open water, and the body size at the shift to this habitat depends on predation risk from largemouth bass. In the vegetation, the young bluegill exerts strong competitive effect on other refuging littoral fish, including the young largemouth bass. As bluegill outnumber the other species, the effect of juvenile bluegill on young of the year bass is larger than the reverse (Olson *et al.*, 1995).

Findings of Hosn and Downing (1994), who studied the influence of cover on the spatial distribution of littoral zone fish in Lac de l'Achigan, Quebec, Canada, which has an average macrophyte biomass, confirmed the hypothesis that prey are less aggregated at sites where predation pressure is low. At low densities, fish in open habitats above the bottom are consistently aggregated. At these low densities, fish in macrophytes above the bottom are nearly randomly distributed. If aggregation is a predator-avoidance response, the authors concluded that fish off the bottom may decrease aggregation when they are found within macrophyte beds. Such flexibility in fish behaviour is important when predation risk is variable. Fish are known to constantly reassess their environment and adjust their behaviour accordingly (Magurran, 1990).

In *Myriophyllum* beds of Lake Guntersville (Alabama, USA), the fish assemblage was dominated by *Lepomis.* The difference in the density of plant did not seem to affect the density of the fish. The canopy of *Myriophyllum* allowed the fish easy access to the underlying water column for feeding while maintaining a relatively high degree of protection from predators (Killgore *et al.*, 1988).

Some predators such as pike, are adapted to catching their prey among vegetation, and other predators, such as largemouth bass (*Micropterus salmoides*), an important sport fish in water bodies of southern USA, may change their predatory tactics from searching to ambushing as plant density increases (Savino and Stein, 1989). Bluegill predation rates decline with increasing plant density (Savino *et al.*, 1992). Young brook trout (*Salvelinus fontinalis*) was found to stay motionless under the cover of vegetation or rocks on the bottom to lessen the risk of predation. By staying near the bottom, fish can also lessen attacks from wading predators.

Most small impoundments used for fisheries purposes in the southeastern USA are stocked with combination of largemouth bass (*Micropterus salmoides*) and bluegill (*Lepomis macrochirus*). Predator-prey relationships between these two species are manipulated by managers to establish a state of balance. Eradication of aquatic macrophytes is a recommended component of fishery management programmes. Where the eradication is either unsuccessful or intentionally neglected and aquatic macrophytes are present, predation on bluegill by largemouth bass declines as cover becomes increasingly complex.

Bluegill is commonly associated with aquatic vegetation in shallow water near shore, but the larger individuals are located in deeper open water. Bluegill is an opportunist, feeding from the surface, off the bottom, or from within the water column. Because bluegills can be easily caught with minimal equipment by inexperienced recreational fishermen, their popularity is widespread and the species has been widely introduced into the western USA, Canada, Europe, South Africa, Japan and Puerto Rico (Martin, 1981). Duffy and Jackson (1994) compared relative abundance, size distribution and condition of small bluegill among stands of three aquatic macrophytes common to lentic environments in the southeastern United States: water lily (*Nymphaea odorata*), water shield (*Braseria schreberi*), and pondweed (*Potamogeton nodosus*).

Nymphaea and *Braseria* are characterised by horizontal, floating leaves, vertical stem, and relatively large interstices between stems, while *Potamogeton* is characterised by dense underwater leaf/stem complexes and relatively small interstices. *Nymphaea* and *Braseria* did not differ with respect to CPUE for bluegill number nor weight. Significantly longer and heavier bluegill were captured from *Potamogeton*, but fish condition was the same for all the plants. Selective management for plants such as *Potamogeton*, which have dense, underwater leaf/stem complexes and relatively small interstices, may reduce the predation by largemouth bass and assist in the maintenance of viable bluegill stocks. Apparently, *Potamogeton nodosus* decreased capture success of larger, more mobile juvenile fish, with the predators switching to alternative invertebrate prey that was less mobile and easier to capture (Dibble and Harrel, 1995). Enjoying such plant protection bluegill may escape predation in greater numbers and at greater sizes. Also, zooplankton abundance and biomass associated with pondweed were greater than those associated with water lily and water shield, thus enhancing the food supply for bluegill.

For Lake Krankesjon in southern Sweden, Hargeby *et al.* (1994) summarized the food web structure before and after the shift from phytoplankton to submersed macrophytes. It shows the favourable effect of macrophytes on the piscivorous pike and large perch, as reported by a commercial fisherman. The favourable effect of macrophytes on the recruitment of piscivorous perch is shown by the increasing size of perch during the plant expansion in the lake. It is probably a result of habitat changes that favour perch before competing cyprinids (Diehl, 1988) and allows a larger portion of perch to shift from feeding on macroinvertebrates to fish. Once a fish assemblage with piscivorous perch and pike is established, this favours the maintenance of abundant submersed macrophytes as the piscivores keep in check the zooplankton-feeding and bottom-feeding cyprinids (Scheffer, 1990).

Jacobsen and Perrow (1998) showed the use of aquatic macrophytes as refuge for roach in a shallow Danish lake: some roach (13%) migrated out from among macrophytes in the absence of predators (pike and large perch) during the day, but during the night roach was out in force (90% of the individuals). When pike was present, roach selected the open water (90–92%) during daylight hours, but kept a 1 m distance from the macrophyte edge. The presence of pike thus reduced the use of macrophytes by roach, which in turn may improve macrophytes and the edge area as a refuge for zooplankton. Adult perch, a generally less effective predator than pike, were concealed within the macrophytes most of the time. Jacobsen *et al.* (1995) pointed out that the potential indirect effect of piscivores reducing the predation pressure upon grazing zooplankton through behavioural changes of zooplanktivores may be an important mechanism in enhancing the stability of submersed macrophytes in shallow lakes.

In a study of the influence of macrophytes on predation in Lake Onalasca, which is a backwater lake of the Mississippi River, age-0 largemouth bass predation on bluegills (*Lepomis macrochirus*) appears to influence the use of vegetated habitat by bluegills. In summer, when most age-0 bluegills were vulnerable to predation by age-0 largemouth bass, bluegill abundance was strongly correlated with vegetation biomass (Dewey *et al.*, 1997).

Sondergaard *et al.* (1997), based on the results of their experiments on a eutrophic Lake Lyng, Denmark, suggested that pike stocking can be used as a lake restoration tool to increase lake water transparency by creating a trophic cascade. The effect of stocking, however, seems to last only during the season in which it has been undertaken, the impact being most significant at high stocking densities. The method can be used in shallow, turbid lakes in which the nutrient loading has been sufficiently reduced to allow a substantial and permanent macrophyte coverage if clearwater conditions are established.

Diehl and Kornijow (1998), who studied the influence of submersed macroinvertebrates on trophic interactions among fish and macroinvertebrates, pointed out that many macroinvertebrate-feeding fish are vulnerable to piscivores and use littoral vegetation as a predation refuge. Consequently, the presence of submersed macrophytes in lakes create the opportunity for different classes of macroinvertebrate-feeding fish to segregate by habitat.

In the presence of piscivores, small vulnerable classes tend to seek refuge in vegetation. The effects of submersed vegetation on pattern of habitat use and individual growth of fish are likely to feed back on the population dynamics of the fish. For example, vulnerable stages of bluegill sunfish in lakes of north America attain high individual growth rates in the open water when vulnerable stages are restricted to the vegetation by piscivorous largemouth bass. The authors also noted that in temperate European lakes the abundance of piscivores is correlated with the abundance of submersed macrophytes.

INVASIVE AND NUISANCE AQUATIC PLANTS AND FISH

Massive occurrence of floating aquatic plants is a common feature on tropical and subtropical water bodies, and it is commonly the result of an expansive growth of non-endemic aquatic macrophytes. In a 1989 report to FAO, Hartley (CSIRO, Brisbane, Australia) estimated that two million people in Nigeria, including more than 24 000 fishermen, and approximately 100 000 persons living in riverine communities in Benin, who rely solely on fishing for their livelihood, are negatively affected by floating aquatic weeds (water hyacinth, salvinia, and water lettuce - *Pistia stratiotes*).

Fishermen in Malawi and Ghana may be hampered by water lettuce. In the Niger, the more recent explosive growth of water hyacinth has interfered with fishing on floodplains of the Niger River. In lakes Kyoga and Victoria in East Africa, until very recently, thousands of fishermen were trying to keep open access to landing and fishing sites blocked by water hyacinth. Water hyacinth and salvinia are also widespread in the tropical zone of Asia and Latin America, representing a major problem for fisheries, irrigation, hydropower production and water transport.

Both the floating and submersed aquatic macrophytes can become nuisance (noxious) plants, especially where they appear as exotics. The problematic submersed plants are: *Myriophyllum spicatum, Hydrilla verticillata, Ceratophyllum demersum, Potamogeton crispus, Potamogeton pectinatus.* The reason for their explosive growth is often the favourable nutrient environment, and also human impact through eutrophication of some, especially small water bodies.

The rapid expansion of the exotic water hyacinth in Sudan reduced the number of fish and fishing intensity (Davies, 1958/59). The periodical annual invasions of the Kerala (India) backwaters with salvinia renders fishing and boating impossible during certain months (Thomas, 1979). In the Sepik River in Papua New Guinea, where in the late 1970s and early 1980s salvinia spread over numerous floodplain lakes, eventually covering over 200 km^2, the tilapia (*Oreochromis mossambicus*) fishery collapsed (Mitchell *et al.*, 1980). The disruption of traditional subsistence lifestyle became so great that entire villages were abandoned (Anon., 1985). Fortunatelly, biological control using the beetle *Cyrtobagous salviniae*, which was released over the infested areas, was so effective that within only a few years salvinia virtually disappeared from all backwaters.

Not always the presence of invasive plants becomes a nuisance. On Rawa Pening reservoir in Java, Indonesia, where all fish species are captured for food, control of water hyacinth led to overfishing, and only expansion of aquatic weeds due to the neglect of control measures led to rehabilitation of fish stocks (Carlander, 1980). Later on, a certain balánce has been achieved among the water hyacinth area cover, capture fishery, and fish cage fish production.

The impact of invasive plants on fish stocks and fisheries is poorly documented, especially for the tropical and subtropical countries. Long-term monitoring is mostly lacking. Records available for the freshwater Lake Naivasha in Kenya, are an example of well-documented changes in a lake ecosystem under the impact of changes in aquatic macrophytes, but also of the impact of human interference with this freshwater lake.

Over the last 30 years, Lake Naivasha, situated at an altitude of 1 890 m, has experienced considerable changes in its plant as well as animal communities. *Salvinia molesta* appeared in the early 1960s, and toward the end of the 1960s it covered 2–3 km^2. Its appearance, combined with heavy fishing pressure on tilapia stocks, resulted in a decline to a level that the local frozen fillet plant had to be closed (Gaudet, 1976). Dissolved oxygen concentration underneath the mats was only 10 % saturation as compared to 64–85 % in the open water (Kongere, 1979). On Lake Kariba, Donnelly (1969) observed that tilapias established nurseries in shallow water only if it was free of *Salvinia*. Where salvinia mats shaded the water, phytoplankton production declined and the young tilapias moved out of the shallows into the deepwater *Potamogeton* and *Ceratophyllum* beds, where they were much more susceptible to predation. In Lake Naivasha the large mat of *Salvinia* almost disappeared by 1987, when it became limited to the river inflow along the northern shore (Harper *et al.*, 1990) Since 1991 the introduction of the biological control agent, the beetle *Cyrtobagous salviniae*, has kept salvinia under control (Abiya, 1996).

The Lake Naivasha level fluctuations, with a high level lasting from 1960 to 1976, followed by a drop, which was followed in 1982 by a sharp rise in water level, has had considerable impact on other aquatic plants. In 1983 the water level started to decline again and by 1985 Kenya experienced a severe drought. By 1987 the lake level returned to the state of the 1950s. Papyrus (*Cyperus papyrus*) plants, formerly floating on the surface, became restricted to margins, but water-lilies (*Nymphaea caerulea*) started germinating, and extensive beds of submersed plants such as *Ceratophyllum demersum* and *Potamogeton* spp reappeared.

The impact of these changes on fish and crayfish revealed that *Tilapia zillii* and *Oreochromis leucostictus* (both introduced species) showed resistance to the changes in lake level fluctuations and subsequent changes in aquatic plants. Both fish feed largely on detritus, which has always been in abundant supply. In *T. zillii* macrophytes constituted less than 10% of the total food taken, while there were none eaten by *O. leucostictus*. Decline in tilapia fishery resulted from intensification of the fishery and the use of small-meshed nets. But whenever the lake water level rose, the flooding of terrestrial vegetation and shallows provided good nursery areas and this led to the recovery of fish stocks (Abiya, 1996). Crayfish *Procambarus clarkii*, introduced in the lake in 1970 and another important component of the lake fishery, is a major consumer of aquatic plants. In the early 1980s it reached a density of about 4 crayfish m^{-2} in the littoral (Muchiri *et al.*, 1995). The species is known to reduce aquatic vegetation in lakes in the USA at densities over 3 m^{-2} (Magnuson *et al.*, 1975). The crayfish fishery in Lake Naivasha crashed in 1983, this being followed by the recovery of rooted aquatics. This provides a circumstantial evidence that the decline and subsequent recovery of the native rooted submersed and floating aquatic plants was due to the heavy crayfish herbivory until 1983 and its relaxation thereafter (Harper, 1992). By 1987 the crayfish density was only 0.25 ind. m^{-2}.

Eichhornia Crassipes - Water Hyacinth

Water hyacinth is now widely spread throughout tropical and subtropical countries. Gopal and Sharma (1981) reviewed the available information on this plant for the years preceding their publication. They also provided maps on the global water hyacinth distribution to that date.

Rzoska (1976) described the invasion of *Eichhornia crassipes* in the Sudanese White Nile. The upper Nile swamps (the Sudd) have become a 'perennial' centre of infestation for the downstream Nile. In many water bodies of the Nile floodplains water hyacinth has fringed the shores in a belt of variable width, and it has taken over the function of *Pistia.* In fringing plants, up to a width of about 6 m, Bailey and Litterick (1993) found a rich community of aquatic macroinvertebrates inhabiting the root of this floating plant, and Hickley and Bailey (1986, 1987) found rich fish communities in water hyacinth fringes. Similarly, Green *et al.* (1976) found also rich fish fauna in water hyacinth fringes of Lake Lamorgan (Java, Indonesia).

However, Davies (1958/59) and Rzoska (1976) noted that fisheries were impeded, specifically in areas inhabited by Shilluks, immediately north of the Sudd in Sudan, and on the Sobat River, also in Sudan. Gopal and Sharma (1981) estimated that in the 1950s about 45 000 t of fish were lost in West Bengal (India) due to the infestation of waters with water hyacinth. Referring to literature, they reported that fish production declined from 905 kg ha^{-1} at zero percent cover to 281 kg ha^{-1} at 25% cover, possibly due to low densities of phytoplankton as a result of shading by water hyacinth, and also the uptake of phosphorus by this plant. One observation reported suppression of the phytoplankton production underneath 10% and 15% cover to 2.25 g m^{-2} and 1.97 g m^{-2}, respectively, as compared to 2.58 g m^{-2} at 5% cover (Gopal and Sharma, 1981). In experimental ponds in Alabama the production of the phytoplankton-feeding *Oreochromis aureus* was reduced by about 50% at 10–25% water cover by water hyacinth (McVea and Boyd, 1975).

In other situations, thick mats of water hyacinth cause complete depletion of dissolved oxyxgen content and consequently fish kills. The presence of water hyacinth in shallows makes such areas unsuitable for fish spawning. In the Sudd of Sudan, Bishai (1960/61) measured 1.8 mg L^{-1} of dissolved oxygen 30 cm below the mat when there was no water current, and he also reported high CO_2 concentrations, which increase the susceptibility of fish to low dissolved oxygen concentrations. Poi de Neiff and Solis (1994), who measured dissolved oxygen concentrations among the roots of water hyacinth floating islands on the Parana River floodplain in Argentina, recorded a maximum of 2.3 mg L^{-1} within the first 1 m depth, but the usual concentrations were close to 1 mg L^{-1}, declining with depth (Fig. 28).

Two Amazon River floodplain fish species, *Colossoma macropomum* and *Schizodon fasciatum*, have shown considerable resilience to hypoxic conditions. Experimental investigations demonstrated that if the hypoxic conditions persist, these species return to the zones of plant cover and remain among the roots of floating plants, where the oxygen availability among the roots seems to be better than in the deeper water. In experiments with *Eichhornia crassipes* and *Pistia stratiotes*, conducted by Jedicke *et al.* (1989), the roots of these plants released oxygen into the water. The input of oxygen beneath a layer of *E. crassipes* was nearly double that beneath an open water surface. The input from water lettuce (*P. stratiotes*), however, was only slightly greater than that determined in open water. Jedicke *et al.* (1989) has pointed out the advantages for the hypoxia-tolerant fish species to remain among the plant roots: reduced predatory pressure; utilization of food supply when other species cannot access it; lower danger to fish from birds and other terrestrial predators, to which fish species displaying emergency respiratory behaviour at the surface of open water zone are exposed.

Bailey and Litterick (1993) found markedly less aquatic macroinvertebrates inhabiting the roots of water hyacinth when appearing as small floating mats, as compared with those of the fixed fringe, and attributed this reduction partly to the feeding activity of fish. It is probable that if the plant appears in broken small-sized mats, which allow to maintain dissolved oxygen concentration underneath the mat at reasonably high levels, its is sought by fish to feed on its periphyton and the associated aquatic invertebrates. The plant also provides a habitat to invertebrate hosts and vectors of parasitic diseases such as malaria, schistosomiasis, encephalitis and elephantiasis.

In the 1970s and 1980s in East Africa water hyacinth invaded the lakes Kyoga, Kwania and Victoria. In Lake Victoria water hyacinth was noticed for the first time in 1988, and by 1992 it was already well established (Ogutu-Ohwayo, 1995). In the 1990s it has been recorded also from Lake Albert. It has had a major negative impact on fish catches and fish landings, due to its obstruction of fishing and landing sites. It is also believed that its presence along the shores has had a major impact on tilapia nesting, through lowering the dissolved oxygen above such sites and shading.

In the early 1990s the worst affected areas have been on Lake Kyoga (FAO, 1993a), but by 1996, in the Uganda sector of Lake Victoria, the weed extended an average of 15 m from the shore, covering almost 80% of the country's 1 000 km of shoreline. Near Kampala, the capital of Uganda, the weed was 30 m wide. Prevailing winds from the south pushed most of the weeds to Uganda, but in 1996 the weed almost encircled the 3 000 km of shoreline (Anderson, 1996). Luckily, the situation has dramatically changed for the better, and by January 2000 it was reported that water hyacinth had been reduced by 80% on Lake Kyoga and 70% on Lake Victoria (Wendo and Ngatya, 2000).

In Lake Victoria only 16% of the shoreline remains infested. This reduction is believed to be the result of the success of biological control using two weevil species, i.e. *Neochetina bruchi* and *N. eichhorniae*, released on Lake Victoria since 1996, and on Lake Kyoga since 1994.

With the decimation of haplochromine stocks by Nile perch (*Lates niloticus*), especially in Lake Victoria, a potential benefit resulting from the presence of water hyacinth has been observed: this floating plant appears to provide cover for haplochromine stocks which have been on the increase since the appearance of water hyacinth in Lake Kyoga. In the reservoir Rawa Pening in Java, Goeltenboth and Kristyanto (1987) observed that water hyacinth is a valuable habitat for fish as long as it does not form extensive solid mats.

Ogutu-Ohwanyo (1995) in his comments on the potential for controlling water hyacinth in Lake Kyoga implies that reduction in the water hyacinth cover might again expose haplochromines to Nile perch predation. The haplochromine-aquatic macrophyte relationships are far from clear. Witte *et al.* (1992) suggested that continuous digging by haplochromines in Butimba Bay, Lake Victoria, could have prevented the settling of rooted plants such as *Nymphaea* spp. With the decline in haplochromines, since the end of the 1980s, the shallows in this bay have been overgrown by *Nymphaea* spp, *Ceratophyllum demersum* and invaded by water hyacinth. Witte *et al.* (1995) believed that water hyacinth invasion of Lake Victoria threatened haplochromines and other fish. It would be interesting to know the result of the presently rapidly diminishing water hyacinth cover on haplochromines and other littoral fish.

In Ghana, water hyacinth was found for the first time in Tano lagoon, apparently entering this water body from Cote d'Ivoire. More recently fishermen reported a decline in fish catches and it was suggested that this was due to the presence of the water hyacinth. Much of the water hyacinth in Tano lagoon is mobile, being moved by wind or water currents. Holcik (1995), who investigated the situation, thought that the low catches are not due to the decline in fish stocks but rather a result of both the inefficient deployment of nets and of the difficult access to the fishing grounds due to the water hyacinth cover.

In Laguna de Bay, a large lake near Manila, the capital of the Philippines, infestation by water hyacinth has been a major environmental problem in several areas of this lake. Massive proliferation of the plant not only poses a hazard to navigation, but it also leads to poor water quality in this shallow lake, which has considerable fisheries importance. Tilapia and milkfish have been grown in numerous fish pens, the area of which at its maximum expansion reached almost 30 000 ha, representing about one third of the total water surface. Frequent typhoons have been known to drive the floating plants against fish pens, destroying them and releasing the penned fish into the lake. In the late 1980s water hyacinth cover was estimated to be 1 800 ha, and the plant was also present in tributary rivers, creeks and irrigation canals. This resulted in difficult, if not impossible access to fish pens, fishing sites, damage to fish pens, fishing nets and other fishing gear, reduction in fish population due to the low dissolved oxygen levels and increased carbon dioxide concentrations, blockage of water intakes of irrigation pumps, and interference with water transport. The removal of water hyacinth was largely manual, plus using one plant harvester. Monitoring of water hyacinth infestation in Laguna de Bay was done with the help of remote sensing using Landsat imageries (Bina *et al.*, 1978).

Tin-mine lakes, of which about 4 300 exist in Malaysia, are a consequence of abandoned tinmining pits, which have filled with water. Some of them have been used for aquaculture, and these represent 18.9% of the total area used for freshwater fish pond culture in peninsular Malaysia (Arumugam, 1994). While a total of 55 species of aquatic macrophytes have been identified from these lakes, in such lakes which receive a significant nutrient input, drastic changes in macrophyte and other communities take place. Once fish culture has been abandoned, in those lakes which are used as dumping sites for organic waste, *Eichhornia crassipes* may cover up to 90% of the surface area. Such lakes contain mainly air-breathing fish (*Channa* spp, *Trichogaster pectoralis*) which tolerate low levels of dissolved oxygen underneath the mat of floating vegetation. Lakes lacking the dense cover of floating macrophytes support indigenous fish and tilapias.

The widespread abundance of water hyacinth has attracted the attention of fish feed specialists and fishery managers. Water hyacinth has been tested both as a direct food of fish, and as a fish feed component. For example, in experiments carried out in Java, Indonesia, up to 10% of the feed was either composted or non-composted water hyacinth. It was noted that the feed, fed to *Puntius javanicus* and *Cyprinus carpio*, did not slow down the growth rate of these fish. *Trichogaster pectoralis* has shown a significantly slower growth rate when fed higher levels of substitution (Hutabarat *et al.*, 1986). Numerous other experiments have been performed on the use of water hyacinth as an additive to fish feed, but they are not a subject of this review.

In Indonesia, water hyacinth, together with some other floating plants, is used as a fish attracting device. In shallow lakes, an enclosure made of bamboo poles pushed into the bottom mud, and filled with aquatic plants, is called *'bungka todo'*.

Tree branches or brush are often added to the aquatic plants. In Kalimantan, such systems are used on inland lakes of the Mahakam River in the east. On the Kapuas River in the west, water hyacinth is rarely found as it does not grow well in black waters of inland floodplain lakes, therefore any water hyacinth which is found is treasured and used as shade to protect fish in cages from direct sunshine, rather than using it as *bungka todo* (Petr, 1993).

The plants provide the fish with shade and food, especially periphyton and its associated invertebrate fauna. The floating aquatic plant enclosures are regularly fished. It has been suggested that in lakes threatened by overfishing, such as Lake Tempe in Sulawesi, *bungka todo* could be an efficient way of fish stock protection and preservation in areas set aside and designed strictly off-limits for fishery. From them, the unprotected areas of the lake would then be naturally recolonized, at least by some fish species.

Pistia Stratiotes

Pistia stratiotes is a free floating common aquatic plant in tropical Africa, where it occasionally forms dense and large mats. In most situations, such mats are not permanent, and are moved or dispersed by wind. In some water bodies, *Pistia* may become part of sudd. The plant consists of a rosette of leaves covered with hydrofobic hairs extending above the water surface, and of submersed, finely divided roots reaching to a depth of 15–20 cm. The roots usually harbour a rich community of invertebrate fauna, which in turn is preyed upon by fish and aquatic predatory invertebrates, such as dragonfly nymphs.

In Volta Lake (a reservoir in Ghana, West Africa), large floating mats of *Pistia stratiotes* developed soon after the reservoir started filling. It became abundant especially where shelter against wind and waves was provided by floating or stranded trunks of dead trees, and inbetween dead trees standing in shallows. In places it was knitted together with *Scirpus cubensis*. The biggest expansion of *Pistia* carpets occurred toward the start of rising water level. After the water level stabilized, *Pistia* became less numerous, with large plants gone, and only small plants remaining. In *Pistia* plants, situated well inside a mat, roots contained between 5 000 and 16 000 m^{-2} aquatic macroinvertebrates (5–35 g m^{-2} wet weight), excluding zooplankton (Petr, 1968).

The values are lower than those for the margins of floating meadows of varzea, where Junk (1973) recorded a maximum of 780 000 individuals (62 g m^{-2}). Junk included the very abundant crustaceans (Copepoda and Cladocera) in his samples. Throughout the observation period, in Lake Volta the *Pistia* roots macroinvertebrate fauna was dominated by the predatory dragonfly nymphs, which although low in numbers exceeded browsers and filter feeders in biomass. The presence of macroinvertebrates was considered to be of importance for fish feeding on periphyton and on it associated invertebrates. In 1965, the first year of the reservoir filling, *Pistia* of Lake Volta also harboured the snail *Bulinus forskalii*, which was very numerous at the peak of the dry season, when the water level was slowly declining and the surface water temperatures were at their highest.

In 1966 the first *Bulinus globosus rohlfsi*, a host of schistosome (*Schistosoma haematobium*) parasites, was recorded from the lake (Paperna, 1970) while *B. forskalii* became rare. In one area Paperna collected over 50 *B. t. rohlfsi* from a single *Pistia* plant. When in some areas of the lake *Pistia* started disappearing, the snails thrived particularly on *Ceratophyllum demersum*, which became very common. This change has led to a dramatic increase in the incidence of schistosomiasis among the fisherfolk on this reservoir.

Little is know about the fish feeding on the aquatic fauna and periphyton associated with the submersed parts of *Pistia*. In Lake George, Uganda, experimental catches within and alongside drifting mats of *Pistia*, which entered a bay surrounded by papyrus, haplochromids represented 98.3% of the total number, and 98.2% of the total biomass of fish captured during a 24 hour period (Petr, unpublished). Among the haplochromids *Haplochromis nigripinnis*, *H. aeneocolor*, and *H. angustifrons* represented 66.5, 17.8, and 13.2% by number of the total, respectively, and a biomass in a similar proportion. *H. nigripinnis* is an omnivore feeding predominantly on algae, insect larvae and emerging insects; *H. aeneocolor*, which is abundant along the papyrus edge, feeds on insect larvae and adult aquatic insects, and the guts contain also plant fragments, suggesting that this species may selectively forage on the food among the roots of papyrus and *Pistia* and that the plant fragments are roots taken incidentally with the insects; *H. angustifrons* is predominantly an open lake species feeding on insect larvae, especially chaoborids, and on zooplankton (Dunn, 1975).

Pistia may be used as a shading plant in aquaculture. In tropical Australia, it has been used in a raceway culture of juvenile growout of freshwater crayfish *Astacus quadricarinatus.* The use of this plant was considered advantageous for several reasons: provision of suitable habitat/substrate for juvenile crayfish; photosynthetic production of oxygen; stabilization of water temperature; absorption of nitrogenous wastes; provision of natural food organisms associated with the plant, i.e. epibionts; shading effect; high carrying capacity for crayfish juveniles with reduced intra-specific competition; enhancing survival and growth of the juveniles (Jones, 1988).

Salvinia Molesta

Exotic aquatic fern, the floating *Salvinia molesta* is spread widely through the tropics. Thick mats of this plant prevent or make difficult boat traffic, impede or make fishing impossible, and the plants harbour also invertebrate hosts of water-borne diseases.

Solid mats of salvinia depress dissolved oxygen concentration, increase concentration of carbon dioxide, and occasionally may result in formation of hydrogen sulfide in water underneath it. Lake Kariba, a large reservoir on the Zambezi River, was invaded by salvinia shortly after the formation of this reservoir. After some years salvinia, covering at its largest expansion in 1962 approximately 1 000 km^2, i.e. 23% of the reservoir surface area (Karenge and Kolding, 1995), declined in the 1970s to about 1% (Mitchell and Rose, 1979), largely due to nutrient deficiencies. Salvinia was replaced by increasing amounts of rooted macrophytes down to a depth of around 10 m. In parallel with that there was an increase in the benthic macroinvertebrates and in benthic-feeding fish species such as *Mormyrus longirostris*, *Synodontis zambezensis*, *Serranochromis codringtonii*, *Serranochromis macrocephalus* and *Labeo altivelis*. Increased diversity within the inshore fish community in Lake Kariba was attributed to the development of a mosaic of new niches among the aquatic macrophytes, periphyton and submersed trees where a variety of benthic gastropods, molluscs, insects and other invertebrates provided a new food resource to fish (Karenge and Kolding, 1995).

The lake littoral is, however, not a stable environment due to a drawdown of this reservoir, which serves the production of hydroelectric power for Zambia and Zimbabwe. It was recommended that to promote full development of macrophytes, the drawdown should not exceed 3.5 m. This limit is sometimes exceeded in Lake Kariba, where dry years may irregularly alternate with wet years. In spite of that, the development of a macrophyte belt has enhanced the stocks of juveniles and small fish species such as *S. macrocephalus*, *Brycinus imberi* and *S. zambezensis* which have increased during the 1980s (Karenge and Kolding, 1995). The authors also mentioned that in parallel with the increase of the above fish species there was a decline in small-sized *Hydrocynus vittatus*, *Hippopothamyrus discorhynchus* and *Marcusenius macrolepidotus*. The reason for this decline is not understood.

With respect to aquatic macroinvertebrate populations on *Salvinia molesta* as compared to those on water hyacinth, Arunachalam *et al.* (1980) reported a higher diversity from the first plant, where they found 66 species, than on the roots of water hyacinth (about 30 species) from the freshwater lake Veli in India. The authors do not provide information on the biomass of the macroinvertebrates on these two plants, hence it is difficult to assess which plant is more important for fish as a source of invertebrate prey.

In Lake Kariba, salvinia rootlets were found in stomachs of a great variety of fish species of herbivorous, omnivorous and even piscivorous character. In *Tilapia rendalli* they were found in 19.6% of fish stomachs, but represented only 6.7% of the total food taken (Mitchell, 1976). Salvinia rootlets, detritus or fragments occurred also in: 9.3% *Haplochromis codringtoni*, 11.1% *Haplochromis darlingi*, 19.2% *Pseudocrenilabrus philander*, 29.6%, *Eutropius depressirostris*, 8.1% *Clarias gariepinus*, 24.0% *Synodontis zambezensis*, 38.2% *Micralestes acutidens* (where it formed 19.2% of the total content), 12.3% *Alestes lateralis*, 11.1% *Mormyrus longirostris*, 5.0% *Marcusenius macrolepidotus*, 10.2% *Hippopotamyrus discorhynchus*, 8.1% *Mormyrops deliciosus*. In the majority of fish species it would appear that salvinia rootlets and detritus were taken when feeding on aquatic macroinvertebrates and plankton residing among the rootlets. For example, out of the 19 fish species investigated for food, 15 contained the crustaceans *Cyclestheria* and *Caridina*, 18 contained the nymphs of the mayfly *Povilla adusta*, 14 contained chironomid larvae, 13 contained trichopteran larvae.

In Sri Lanka, salvinia is often removed from water reservoirs manually, but biocontrol using the beetle *Cyrtobagous salviniae* is more efficient.

Cyperus Papyrus

Papyrus rhizomes are typically rooted in loose detritus in a floating mat. Papyrus may occur as free-floating or stranded, or forming fringe swamp. Occasionally, the lake-side edge of the fringe papyrus swamp will float if the lake level permits. The extent of papyrus swamplands in East and Central Africa is enormous. In southern Sudan the Upper Nile forms large swamps called the Sudd, which covers approximately 40 000 km^2 (Rzoska, 1974). In Uganda, papyrus encircles completely Lake Kyoga, and is well developed in Lake Victoria basin.

In Lake George, Uganda, where it forms large fringing floating swamps, papyrus flourishes in water with a conductivity of 200 ìS, while in Lake Edward (850 ìS) it is restricted to river mouths (of much lower conductivities) on the Uganda shore of the lake (Lock, 1974). Gaudet (1980), in summarizing the significance of the papyrus for the aquatic fringe ecosystem, pointed out that papyrus swamps supply large amounts of fixed nitrogen to tropical lakes and rivers every year thus allowing an increase in animal and plant production at the swamp edge. Papyrus fringes act to moderate changes along a river system by regulating nutrient flow and recycling. For the most part this is accomplished by production of an organic nutrient output at the expense of an inorganic nutrient input.

Papyrus plants effectively extract dissolved nutrients from tropical river systems, but such nutrients are later exported as organic particulate matter or adsorbed to particles which are carried into a lake by through-flow (Gaudet, 1979). In practical terms, papyrus swamps may be effective in removal of organic nutrients from sewage effluents into these swamps (Gaudet, 1978), since papyrus seems to be quite tolerant to pollution. The organic nutrients released from the organic particulate matter or adsorbed to particles represent an energy source to a very diverse tropical fauna and flora, which in turn supports a profitable shallow water fish industry in many parts of Africa (Gaudet, 1980).

Summarizing the importance of papyrus swamps along the major rivers in Africa, such as Congo and White Nile, Gaudet characterized them as an ecological buffer zone, which moderates changes along the river system by regulating nutrient flow and recycling. For the most part this is accomplished by production of an organic nutrient input at the expense of an inorganic nutrient input. The organic nutrients serve as an energy source to a very diverse tropical fauna and flora, which in turn supports a profitable shallow water fish industry in many parts of Africa. At lake edges papyrus swamps are often sites of intensive fishing and thus a resource worthy conservation.

For air-breathing fish papyrus swamps and other large swampy divides are not likely to be a barrier to dispersal. For water breathers, the use of and dispersal through papyrus swamps is likely to be limited by the availability of oxygen, the open surface area, and efficiency of oxygen uptake in the species (Chapman and Liem, 1995). In a study of a papyrus swamp, which feeds a small river in Uganda, Chapman *et al.* (1998) found peak of dissolved oxygen concentrations associated with biannual flooding, and they suggested that this provides opportunities for fish to use or move through papyrus. However, peak flood values never reached 50% saturation, and thus may pose barriers for some hypoxia intolerant species. At the lake edge of Lake George in Uganda dissolved oxygen showed a rapid decrease moving inside the papyrus belt (Petr, 1973) (see Fig. 22). Chapman *et al.* (1998) reported *Ctenopoma muriei, Clarias liocephalus* and *Nothobranchius* sp. to be found principally in papyrus swamps and other wetlands of the Lake Victoria Basin. Chapman and Liem (1995) found the small cyprinid *Barbus neumayeri* to inhabit the dense interior of papyrus swamps, due to the presence of large gills and the use of aquatic respiration at the air-water interface. Chapman *et al.* (1995) reported that 12 species of East African cichlids are tolerant of low dissolved oxygen concentration and able to use low oxygen habitat along the margins of Lake Victoria as a refugium against the introduced Nile perch.

In lakes with a more pronounced fall in water level, papyrus stands may become drastically reduced. In Lake Naivasha water level declined by just over 4 m between 1980 and 1987, exposing some 6 500 ha of soils, or about 35% of its 1980 area (Harper *et al.*, 1993). During this decline exposed soils were cleared of vegetation, mostly stranded *Cyperus papyrus* and other emergent plants, and put down to agriculture. Harper *et al.* (op. cit.) noted that fringing papyrus is now voluntarily kept by some landowners to keep a buffer zone, which assists in reducing the input of high nutrient loads into the lake. The increasing demand for water is seen as a direct threat to the lake in the future (Abiya, 1996).

Submersed Aquatic Plants

Myriophyllum spicatum, a native Euroasian plant, spread rapidly across North America in the 20th century, displacing native species in some water bodies and often forming dense stands and surface mats. In Currituck Sound in North Carolina, USA, it was detected for the first time in 1965, when it covered 40 ha. In 1966 it covered 3 200 ha, and by 1974, 32 000 ha were infested (Sheldon, 1997). *M. spicatum* can withstand a broad range of conditions: it can grow in oligotrophic to eutrophic waters, in 0.5 m to 8 m depth, in sand, in organic substrate, under pH ranging from 5.4 to 10, in brackish water, in northern temperate lakes and rivers, and in subtropical water bodies.

In 1978 in Lake Pontchartrain estuary, this macrophyte has become a dominant species, with a lower abundance of the native macrophytes *Vallisneria americana* and *Ruppia maritima* present there. Duffy and Baltz (1998), who in the estuary studied the impact of the invasive *Myriophyllum* on fish, found that the highest fish species diversity was in *Vallisneria*, while the lowest in *Ruppia* and *Myriophyllum*. The common fish were significantly more abundant in vegetated areas than on adjacent unvegetated substrate, and total abundances were higher in *Myriophyllum* and *Ruppia* than in *Vallisneria*. The authors suggested that the exotic

Myriophyllum may not have had a detectable influence on fish assemblages or abundances relative to the native macrophytes because the disturbance of the substrate by waves in more open areas prevent it from growing densely enough to strongly alter the original microhabitat. However, in some other water bodies, *M. spicatum* may form thick, impenetrable walls of uniform height and leaf form. When the stem density exceeds 250 stem m^{-2} fish start avoiding the plant (Savino and Stein, 1982). Keast (1984) found a lower fish density in *Myriophyllum* beds than in native aquatic macrophytes, and Dvorak and Best (1982) found lower densities of aquatic macroinvertebrates in this plant, when compared to native plants.

Sloey *et al.* (1997), who studied the distribution of aquatic macroinvertebrates in beds of *M. spicatum* in Fish Lake, Wisconsin, found higher invertebrate biomass, density, and taxa richness in the upper foliated sections of plants versus the lower unfoliated sections. The edges of the beds contained higher biomass, density, and taxa richness than the centre of the bed. They did not study the impact of the differential distribution of macroinvertebrates on fish, but other studies confirmed that the structural complexity of plant species and spatial distribution of plants are among the decisive factors determining which and how many fish are frequenting aquatic macrophyte stands.

Hydrilla verticillata is another invasive submersed species, which is present in numerous water bodies of the Americas, Europe, Africa, Asia and Australia. In some water bodies in the USA this plant has led to a considerable increase in fish stocks and fish species of sport fishery importance, while in other water bodies its presence has required a management response to prevent decline in recreational fishery. Apart from grass carp, for which this plants is a preferred food, other fish have also been used for controlling this plant. In Africa, *Tilapia zillii* and *T. rendalli* are known to feed readily on it, in India *Cirrhinus mrigala* and *Etroplus suratensis*, stocked in nursery tanks with *Channa punctatus*, efficiently controlled this plant (Raj, 1976), and also in India, *Tor tor* and *Puntius* spp feed readily on it. Herbivorous turtles in Bangladesh, and manatee in the USA and some Latin American countries also feed on hydrilla.

SPECIAL FLOATING PLANT COMMUNITIES AND FISH

Sudd and Floating Islands

Sudd is a large area covered in aquatic plants and situated in Sudan, between 5° 30' and 9° 30' N in latitude and between 30° and 32° E in longitude, where it extends from Juba to Lake No. The word "sudd" is derived from the Arabic word "sadd", meaning "block" (Gaudet and Falconer, 1983). The term is now being used widely for large solid floating vegetation islands and mats. The Sudd region in the Sudan has remained unnavigable for many years because large islands of papyrus have floated downstream of the Nile and lodged in channels, effectively blocking local river traffic. In 1980 the swamps of Sudd covered about 16 500 km^2 (Hickley and Bailey, 1986). *Typha dominguensis* covers about three quarters of the total swamp, bordering *Cyperus papyrus* which forms a central core. Water hyacinth (*Eichhornia crassipes*) now forms an almost ubiquitous floating fringe to river channels and lakes in Sudd swamps.

Both before and after the *Eichhornia* invasion marginal vegetation in the Sudd has been shown to support a wealth of small and juvenile fish (Sandon, 1951; Hickley and Bailey, 1986) and rich and diverse macroinvertebrate faunas (Monakov, 1969). According to Bishai (1960/61) the best fishing grounds, giving a maximum annual yield of fish, were those not infested with water hyacinth. The main fishing grounds in the Sudd are in the main water passages, side channels and lagoons, and fishing camps are established there every year. In his review of the Sudd, Rzoska (1974) concluded that vegetation fringe in which *Pistia stratiotes* had a major role, formed a unique, and biologically the richest, habitat in the whole swamp system.

Although *Pistia* has been largely replaced by water hyacinth, Bailey and Litterick (1993) have found that the fringing hyacinth in Lake Wutchung in the Sudd region has taken over the function of *Pistia* in harbouring large quantities of macroinvertebrates, with the highest concentrations present within the first 6 m of the fringe (Fig.34). While the outer fringe was dominated by coleopterans, odonatans, gastropods (including *Bulinus forskalii, B. truncatus, Biomphalaria pfeifferi, B. sudanica*), and the crustacean *Caridina nilotica*, further in the mat coleopterans, hydracarines and gastropods were prominent. In free-floating water hyacinth rafts covering up to 10 m^2 there were over seven times fewer invertebrates, especially a notable reduction in crustaceans and ephemeropterans, suggesting that the macroinvertebrates there were subject to heavy preying by fish due to an easy access to the underside of such floating plant rafts. This agrees with the findings of low biomass of invertebrates on the isolated, off-shore *Ceratophyllum*, stranded on flooded trees of Volta Lake in Ghana (Petr, 1968), suggesting that apart from the increased exposure to predation, the floating plants have little chance of being recolonized by benthic-dwelling invertebrates.

Hickley and Bailey (1986) sampled fish in two major habitats, i.e. flowing channels and a lake situated within the sudd. The greatest diversity of fish (some 60 species) were caught in the lake's papyrus zone. *Eichhornia crassipes*, which forms a floating fringe of variable width in channels and lakes, were found to harbour rich periphyton and communities of aquatic invertebrates which the plant provides with shelter and food. Deeper inside of the *Eichhornia* mat, with a decrease in the concentration of dissolved oxygen reaching 15–20% of saturation level, the fish typically recorded were air-breathers, such as *Polypterus senegalus* and *Ctenopoma* spp (Hickley and Bailey, 1987). Many non-air breathing river and lake fish species were found to move through conduits in the permanent floodplain created by swamps, but the low dissolved oxygen content or hypoxic conditions, high concentrations of CO_2 and ammonia must prevent their lateral dispersal into the swamp.

A: Fringe and shore profiles. Sites 1 = landward zone; Sites 2 = inner middle zone; Sites 3 = outer middle zone; Sites 4 = outer zone.

B: Estimated fresh weight biomasses of hyacinth (kg m^{-2}).

C: Estimated macroinvertebrate numbers (solid bars) and fresh-weight biomasses (stippled bars, g m^{-2}).

In the Sudd, submersed beds of *Naias pectinata, Vallisneria aethiopica* and *Ceratophyllum demersum*, with their rich periphyton, offer abundance of food to *Heterotis, Distichodus* and several tilapias.

Annual flooding of the permanent Sudd swamps inundates large areas of *Oryza longistaminata* grasslands, where a number of fish species enter, such as *Clarias gariepinnus, Polypterus senegalus, Channa obscura,*

Oreochromis niloticus (Hickley and Bailey, 1987a). In addition, large resident and migrant populations of birds are present in the Sudd. Of these about 20 species feed on submersed vegetation in shallow waters, over 100 species consume invertebrates, and some 40 species feed on fish or their fry (Hickley and Bailey, 1987).

Floating islands occur regularly on a number of water bodies in Africa, including major rivers such as the Nile and Congo. They cause a variety of problems, especially for boat transport, water intakes, but also for fishermen whose nets and other fishing gear become entangled. Floating islands made largely of papyrus have interfered with fishing in lakes such as George (Uganda), Jipe (Tanzania-Kenya), Naivasha (Kenya), Chad, Victoria, and Kyoga. Remote sensing has been used to follow the movement of floating islands on Lake Naivasha (Gaudet and Falconer, 1983), using the Landsat satellite data. This has allowed to determine their mobility, including the separation of the movements of papyrus and salvinia. The major movements take place in January-February, under the influence of the changing wind patterns. Prompt communication of the results to the fisheries community would assist in adapting their fishing strategies.

The word 'sudd' has been applied also to floating islands of aquatic macrophytes formed on some man-made lakes in Africa (Gaudet, 1979a). Within several years after the closure of the dam and filling of these large reservoirs sudd started forming. On Volta Lake (Ghana) mats became established in the shallows around drowned trees, with the major plants being *Vossia cuspidata, Leersia hexandra* and *Oryza barthii* (Photo 14). Some southern parts of the lake developed sudd through colonisation of *Pistia stratiotes* by *Scirpus cubensis* and *Typha dominguensis*, with *Utricularia inflexa* fringing the margins (Petr, 1986). Sudd has also developed on the shallow Kpong reservoir situated on the Volta River just downstream of the Volta Lake.

On Lake Kariba, after *Salvinia molesta* developed thick mats, it combined with *Pistia stratiotes* and *Ludwigia* spp to form sudd. For a number of years the sudd impeded passage and prevented fishing activities in some areas of this reservoir. *Salvinia* together with *Ludwigia* supported seven times more invertebrate biomass than did *Ceratophyllum* and *Potamogeton pussillus*. The benthos underneath *Salvinia* mats was suppressed (McLachlan, 1969). On Lake Kainji, a reservoir on the Niger River in Nigeria, small amounts of *Echinochloa* sudd developed in the northern part (Gaudet, 1979a). Studies on the significance of sudd for fish stocks in African man-made lakes are needed.

Sudd formation is also reported for lakes and reservoirs in Asia and the Pacific. In Rawa Pening in Indonesia, an old reservoirs constructed prior to the second World War, up to 60% of the surface area is covered by water hyacinth. In Bung Borapet reservoir, Thailand, an area of several km^2 is covered with vegetation (*Coix aquaticus, Isachne globosa, Leersia hexandra, Phragmites karka* and *Scirpus grossus*), of which the first three plants form thick floating mats which are very resistant to wind and wave action. Junk (1977) found very low concentrations of dissolved oxygen and the presence of hydrogen sulfide underneath the mats, and he suggested such water is useless for fish, as it would be virtually devoid of invertebrate fauna. *Eichhornia crassipes* and *Salvinia cucullata*, also very abundant, drift around the lake together with the floating islands. In Sirinumu reservoir, which flooded both tropical forest and rubber plantation, and now provides drinking water for Port Moresby, the capital city of Papua New Guinea, suddification took place among the flooded trees.

In the Amazon Basin in South America, non-attached sudd formations are called 'floating meadows' (see the following Section).

Varzea - White Waters

White-water rivers, mainly the Solimoes/Amazonas, harbour fish which are an important source of protein for the local population. Varzea floodplains contain a dense aquatic vegetation called 'floating meadows', composed of the grasses *Paspalum repens* and *Echinochloa polystachya*, interspersed with several other plant species. During the high water season lakes within the floodplain are connected by channels to the main rivers and the grasses form floating mats with submersed mass of rhizomes and roots which support a rich invertebrate fauna. These, together with detritus, provide an important food for small fish species and juveniles of important commercial species such as migratory characins. Characins spawn in varzea lakes at the beginning of the flood and their juveniles inhabit and feed among the floating plants. The three common fish of white waters near Manaus. *Schizodon fasciatus, Rhytiodus microlepis* and *R. argenteofuscus*, are herbivorous (Mendes dos Santos, 1981), alongside of a number of other herbivorous and predominantly vegetarian fish species.

The submersed parts of floating meadows are used by the aquatic invertebrates as a substrate and partly for nutrition. In varzea lakes the density of aquatic invertebrates reaches 780 000 individuals m^{-2} (11.6 g m^{-2} dry weight, or 62 g m^{-2} fresh weight) at the edge of the stands (Junk, 1973). Towards the inner parts population density and biomass decrease. In very dense stands anoxia prevails. However, population densities and composition can vary strongly in monthly intervals, in different parts of a lake, and between varzea lakes (Junk and Robertson, 1997). The invertebrate fauna of the floating vegetation of the varzea is considered to be the richest in the Amazon. Under the influence of wind and flow the meadows may break into pieces ranging in size from a small clump to a football field and pass down the lakes to the main river channels (Henderson and Hamilton, 1995).

In Lago Mariraua, an upper Amazonian floodplain lake of varzea type, for anchored meadows, total wet weight of fish was estimated in the month of December as 312 kg ha^{-1} and the number as 90 400 individuals ha^{-1}. In drifting islands, fish biomass and density were estimated as 192 kg ha^{-1} and 12 700 individuals ha^{-1}, respectively (Henderson and Hamilton, 1995). According to Lowe-McConnell (1987) the fish standing crop in tropical fresh waters ranges from 55.7 to 5 116 kg ha^{-1}. In Lago Mariraua the biomass estimate was probably below the annual maximum because it was made when the water was rising and just begun to enter the flood forest.

At this time of the year many pieces of floating meadows are breaking free. Within the canal outflowing from the lake there was a continuous passage of at least 30 islands per day, with a total surface area ranging from 100 to 1 000 m^2. Fish families Cichlidae and Anostomidae favoured marginal, anchored meadows, but were frequently dispersed by drifting islands. The synbranchids and gymnotids (swamp eels and knifefish) are the groups most adapted for life within floating meadows. None of these were caught within the drifting islands. This suggests that these fish are able to detect changes associated with the break-up of the meadows and move into secure areas. The Siluriformes favoured drifting islands over anchored meadows. For this group the downstream drift may represent a seasonal migration to the main river channel. Juvenile fish avoided drifting islands, presumably to avoid being flushed from the lake. It is estimated, that within the drifting islands 14 to 140 cichlids, 9 to 90 anostomids, and 74 to 740 catfish were transported daily out of the lake.

Aquatic macrophytes occurring in whitewater floodplains in large quantities are of poor nutritional value in comparison with other food items (Junk *et al.*, 1997). Stomach contents of 91 species of fish, analysed for food from varzea and igapo (black waters) did not show any fish to be feeding to a major extent on aquatic macrophytes. Wherever possible, the fish prefer high-protein algae with a C:N ratio of about 9.5 and fruit rather than low protein macrophytes with a mean C:N ratio of about 25. Aquatic macrophytes are preferred by juvenile fish not because they are a good food value, but because they offer shelter and a large variety of food items, such as detritus, periphyton, perizoon, and terrestrial invertebrates. As the juveniles become larger, they move to other habitats, such as the open water or the floodplain forest and change their diet. The indirect importance of macrophytes as a food resource for fish, for example, in the form of macrophyte-feeding terrestrial and aquatic invertebrates, detritus and associated bacteria and fungi as well as in the form of dissolved organic carbon from decaying macrophytes via the microbial loop, has not yet been evaluated (Junk *et al.*, 1997).

From the two freshwater streams in Costa Rica (not to be classified as varzea system) investigated for fish, in the larger one fish were found to be closely associated with dense mats of floating aquatic plants *Eichhornia*, *Hydrocotyl*, *Salvinia* and *Azolla.* Three eleotrid fish species *Dormitator maculatus*, *Eleotris pisonis* and *E. amblyopsis* were almost always captured from the dense beds of floating macrophytes that lined the stream margins and sometimes completely covered the channel (Winemiller and Ponwith, 1998). The two *Eleotris* and to a lesser extent *D. maculatus*, were also occasionally captured from submersed stalks of immature water hyacinth. Juvenile *Gobiomorus dormitor*, the fourth eleotrid, were rarely captured in aquatic vegetation, although individuals of all sizes were sometimes captured near or underneath patches of floating plants. *G. dormitor* was the only eleotrid associated with more open habitats and was collected in the channel area. *D. maculatus* feeds mostly on coarse detritus and minor amounts of filamentous algae and diatoms, while both *Eleotris* species and *G. dormitor* were carnivorous, feeding largely on shrimp and small fish. There was a shift in the food as the fish grew bigger. Diet similarity among the three carnivorous eleotrids declined during the wet seasons, the period when their prey disperse at lower per-unit-area densities within the flooded forest that surround the streams.

Igapo - Black Waters

Black and clear water regions of the Amazon Basin also contain floating meadows of grasses (Graminae), but these are less developed and diverse than those of white waters. This may be responsible for the better dissolved oxygen conditions in varzea lakes as compared to blackwater lakes. Araujo-Lima *et al.* (1986) studied the fish composition underneath two floating plants communities, e.g. pure *Oryza* sp., and a mix of *Oryza* sp. and *Paspalum repens*, in a blackwater floodplain lake of the Rio Negro. The main difference between these two microhabitats was the complexity of the submersed root system. A higher concentration of fish was found under mixed vegetation. Younger fish of *Metynnis cf. hypsauchen* fed predominantly on Cladocera, Chironomidae and Trichoptera, switching over to filamentous algae and wild rice seeds. The food of *Moenkhausia lepidura* was similar, but this fish kept insects as food when it grew larger. The third species, *Hemigrammus levis*, fed mainly on Cladocera when young, changing more to chironomid larvae and other insects when larger.

Due to the lower biomass of aquatic plants and phytoplankton in the blackwaters as compared to whitewaters, fish in the Rio Negro depend much more on the flood forest than on the floodplain (Junk *et al.*, 1997). Adult fish tend to overcome food shortages by their ability to feed on diverse items. Catch per unit of effort (gillnet) was on average 210 g m^{-2} day^{-1} in varzea as compared to 41 g m^{-2} day^{-1} in the igapo, and the catches are highest during the falling and low water levels. Due to natural diurnal fluctuations in dissolved oxygen content within and underneath macrophyte stands, most fish regularly leave the aquatic plants during the night and seek out open waters. Most Amazonian fish are resistant to low oxygen concentrations, with 36 out of 43 species tested in experiments able to tolerate concentrations lower than 0.5 mg l^{-1}, at least for some hours (Junk *et al.*, 1997). By about 8 a.m., when the dissolved oxygen content in macrophyte stands has risen as a result of the photosynthetic activity of phytoplankton and periphyton, the fish return.

The vegetation of the islands in igapo is characterized by flood forests. Most of the plant species display marked periods of leaf production and fall. The period of highest litter production is during the flood months, the maximum being in July. The expansion of the littoral zone from the main channel into the forests is accompanied by the expansion of the benthic fauna. During the flood period the highest numbers of shredders and macrocollectors are recorded from the igapo. Macroinvertebrate scrapers are limited to the rising and falling water periods.

In igapo, as on other types of floodplains, the main factor governing the aquatic productivity is the flood pulse (Nessimian *et al.*, 1998). In this environment, the benthic fauna represents perhaps the most effective component in the energy flow between the aquatic and terrestrial compartments. Large river water level fluctuations force the fish to have flexible feeding and other habits. Characoid fish species, with their fast growth, early maturity, and high reproductive rates are well adapted to such changes and represent considerable part of fish biomass in varzea lakes. The blackwater river floodplain of the Rio Negro igapo system, in spite of having a higher fish species diversity, has a low fish biomass, which is only about 20% of that found in the whitewater floodplain of the Amazon River (Junk *et al.*, 1997),

INUNDATED FORESTS, SUBMERSED TREES IN RESERVOIRS AND FISH

Inundated forests, flooded forests, or flood forests are the names given to forested areas periodically flooded during the flood season. They are present largely in the tropical countries of Latin America (e.g. igapo and varzea), in tropical Africa, in tropical Asia, and the island of

New Guinea, but forests along many rivers in subtropical and temperate zones are also adapted to withstand longer term flooding. Inundated forests represent an important habitat for fish, which find among the flooded trees a temporary breeding and nursing habitat as well as rich feeding grounds. Their importance for freshwater capture fisheries has been emphasized in a number of publications.

New reservoirs frequently permanently flood the existing forests as clearing forests on large areas is often uneconomic. Leaves are the first part of trees which undergo decomposition, but also attract herbivorous fish. The decay of the wooden parts is slow, and may last hundreds of years. Trees which are regularly exposed to drawdown, are subjected to the attack of wood-burrowing insects and fungus, and when submersed, attacked by microorganisms, and in some water bodies also by aquatic insects the larvae of which seek shelter in burrows. In the euphotic zone, the submersed surfaces are colonized by epiphyton. Both the epiphyton and the associated micro- and macroinvertebrates are an important food of fish.

Inundated Forests in Asia, Oceania and Latin America, and Fish

In some countries forest trees are resistant to long-term flooding. Such forests are present in the Amazon, in southeastern Asia, Papua New Guinea, and the USA. During high water levels the flooded forests provide habitat and submersed surfaces for periphyton production. In Cambodia, around the Great Lake (Tonle Sap), in 1990, inundated forests and their degraded remnants covered around 500 000 ha. Satellite imagery showed that 361 700 ha were flooded evergreen forest, 152 200 ha a mosaic of flooded forest, swampy vegetation and fallow land.

As a result of human activities around the lake, the forest has been rapidly diminishing, with encroachment of agriculture, paralleled by clearing of the natural vegetation, having the greatest impact. Grazing cattle in the revegetating forest prevents its recovery.

The inundated forest of the Great Lake, and to a lesser extent the inundated forest of the Tonle Sap and Mekong rivers, are considered to be essential for maintaining the existing level of inland fishery production in Cambodia (Csavas *et al.*, 1994). About 90 percent of the total freshwater stocks follow the inundation spawning pattern and many of fish species breed in the inundated forests, with the lake water rising during the peak of floods by 10 to 14 m, expanding the lake surface area from 2 500 km^2 to 11 000 km^2, exceptionally to 13 000 km^2. Since the 1950s the exploitation of inundated forests around the Great Lake has accelerated under the severe pressure for agricultural land development, for charcoal and firewood production, smoking fish, construction of fish traps, for firing brick kilns, and also due to the use of branches for brush parks (fish attraction devices).

As no solid information exists on the significance of these inundated forests on the individual species of fish, it is difficult to assess the damage which the rapid exploitation of the forests has caused to fish stocks. Some areas of the degraded forest are now covered by regrowth of different species with a height up to 4 m. This may have ameliorated the detrimental impact of deforestation on fish populations. A pilot project was initiated in the 1990s on the rehabilitation of some parts of the forest, aiming at replanting water-tolerant trees. The results are yet to be fully evaluated.

In Lake Luar, the largest lake on the middle Kapuas River in Indonesia, close to the Sarawak border, the presence of inundated forests is believed to sustain a high fish production in the otherwise nutrient impoverished black water (Petr, 1993).

In the lowlands of the Sepik and Ramu rivers in Papua New Guinea more than two thirds of the estimated 33 000 km^2 of floodplains are covered by seasonally inundated floodplain forest. In addition to that, at altitudes below 200 m there is an estimated 55 000 km of permanent river channel wider than 2.5 m, the majority of which is bordered by dense forest, flooding in the rainy season at altitudes below 50 m (Coates, 1993). Inputs of vegetable matter, especially fruits, nuts and other seeds into rivers, lakes and flooded forest from this large expanse of terrestrial vegetation are considered to be relatively high, and Coates believes that the flooded forest in Papua New Guinea broadly supplies an equivalent quantity of plant products as produced by flood forests of the Amazon. The only indigenous species utilising this resource is the halfbeak *Zenarchopterus kampeni* which feeds on very small seeds, but mainly on insects, falling onto the water surface from surrounding vegetation (Coates and Zwieten, 1992). It was therefore decided to introduce from Malaysia the fruit-eating south American fish species *Colossoma bidens* (=*Piaractus brachypomum*). The introductions were made between 1995 and 1997 (Coates, 1997). The results of the introduction are not known.

In the Brazilian Amazon, the temporarily inundated forests extend over an area of some 70 000 km^2 (Pires, 1972). In a study by Waldhoff *et al.* (1996) on white water (varzea) and black water (igapo) 23 fish species (22% of species captured in the area) from inundated forests fed on fruits and seeds in varzea, and 34 species (13% of species) on the same in black waters. Whitewater and blackwater rivers are not completely distinct entities with separated fish faunas, since individual fish of many species move from one type of water to another seasonally and during the course of their life history (Lowe-McConnell, 1987)

Altogether, in the Amazonia about 200 fish species of fish feed on fruits and seeds of the seasonally inundated forests. The most common of them is tambaqui (*Colossoma macropomum*), a good table fish, which feeds almost entirely on fruits and seeds of the inundated forests during the highwater period, while starving in the main river channel during low water. Food and seed production in the inundated forests is much higher than in other forests (Revilla, 1991). Waldhoff *et al.* (1996) estimate that of the total of 3.5 billion trees in the inundated forests in Amazonia about a quarter produce important fish food every year during high water. If only 10% of this production is consumed by fish, and assuming a conversion factor of 1:10, the fruit output corresponds to a commercial fish production of between 160 000 and 530 000 tons y^{-1}.

The floodplains in the Central Amazon Region close to cities and towns are now intensively used for agriculture. The varzea forest has been removed and large areas are used for grazing (Conserva and Piedade, 1998). Bayley (1989) reported 50% of the forest near Manaus has been already cleared. Deforestation rate is larger further downstream. Araujo-Lima *et al.* (1998) estimated the economic value of *Colossoma macropomum* for the economy of Manaus, the largest city in Central Amazon. In 1993–1994 it was estimated at US$ 8.3 million. Flooded forest revenue related to tambaqui production was larger than poultry and rubber production in the Amazonas State, and close to the value recorded for the logging industry. Araujo-Lima *et al.* concluded that conservation of the flooded forest habitats is of critical importance for the future of the local economy.

Submersed Trees in Reservoirs and Fish

While inundated herbaceous plants in a new reservoir decompose usually within a year, submersed trees may remain intact for over 100 years, especially those that are never exposed to the atmosphere (Jenkins, 1970). Submersed and partly emerged trees and brush provide surface area for periphyton and the associate invertebrate fauna. In some reservoirs, the primary production of periphyton may exceed that of plankton, but exact data are not yet available. Where dense enough as well as where they are standing in shallow water, trees also provide shelter for the establishment of macrophytes which they protect from wave action. The habit of some fish species to deposit eggs on underwater surfaces provides such species with an enlarged suitable breeding and nursery habitat. Many centrarchids and some other fish of North America have been observed to build nests on inundated herbaceous vegetation, trees, brush and stumps (Ploskey, 1985). For example, the abundance of yellow perch (*Perca flavescens*) was directly correlated with the availability of brush. Davis and Hughes (1971) found that in Bussey Lake, a reservoir in Louisiana, large bluegills and largemouth bass were more abundant in areas with trees than in those without trees. Reservoir coves with submersed trees also have a higher fish standing crop than coves without inundated trees, as found in a study of Willis and Jones (1984) in seven Kansas reservoirs.

In Parakrama Samudra reservoir in Sri Lanka, prior to submergence, large areas were covered by trees. These are now in various degrees of degeneration. Although the trees represent a danger to fishing gear, fishermen prefer setting the nets among trees as fish tend to congregate in such areas and this makes commercial fishing viable. De Silva (1988) suggested to leave approximately 50 % of the forest cover of the future reservoir bed in new reservoirs, established in forested areas. The retention of some uncleared areas for fisheries development in new reservoirs has also been supported by Ploskey (1985), who evaluated the effect of preimpoundment clearing on reservoir ecology and fisheries in North America.

In the Volta Lake in Ghana, West Africa, much of the 8 480 km^2 flooded by this reservoir was originally tree-covered.

After flooding, trees in the shallows were rapidly colonized by the mayfly *Povilla adusta*, the nymphs of which burrow into the bark and in soft wood tree species (Petr, 1970, 1986a). The filling of the reservoir extended over a period of four years. As the water was advancing and rising, the *Povilla* nymphs attacked the newly submersed trees. Of the littoral plants which later appeared, they preferred the hollow-stemmed *Polygonum senegalense*, which provided readily available protective cavities. During the night the nymphs left the cavities to filter-feed on planktonic algae and graze periphyton on the surface of the dead trees and aquatic macrophytes. A primary consumer, *Povilla* became an important link in the food chain between algae and fish, and its mass occurrence has been conducive to the maintenance of many fish stocks, especially the insectivorous *Alestes*, *Eutropius* and *Schilbe*, which in 1967 constituted 81% of fish landings by number in the north of the reservoir. Stomachs of the characid *Brycinus leuciscus* also contained a high proportion of *Povilla*.

Mormyrids *Mormyrus macrophthalmus* and *Mormyrops deliciosus* preferred the big nymphs of *Povilla* and dragonflies to chironomid larvae, the first fish species largely being limited to the riverine reach of the reservoir (Petr, 1968a). When *Povilla* became most abundant in the reservoir, i.e. in the third year after flooding (1967), in the characid *Alestes baremoze* it was present in 65% of the stomachs, representing 56% of the total food eaten (Petr, 1974). *Povilla* was also the dominant food of some *Brycinus macrolepidotus*, but more commonly this species was omnivorous, with plant material a regular component of food in its stomach during the rising reservoir water level. The pelagic fish *Pellonula leonensis* and *Physailia pellucida* also fed on this mayfly, which represented the major food item for the first fish (Reynolds, 1970). Altogether 14 species of fish have been found to feed on *Povilla* in Volta Lake. With the full development of *Povilla* on/in inundated trees of Lake Volta, there appeared to be no indication of any interspecific or intergeneric competition for food among the omnivores and predators (Petr, 1974).

The fresh mass of *Povilla* nymphs from trees ranged from 6–69 g m^{-2} of bark. This is in excess of the zoobenthic maximum standing crop (wet weight) of 6.1 g m^{-2} found in the most productive part of Volta Lake. In Lake Kariba, when comparing the aquatic invertebrates of inundated trees with those of benthos and aquatic vegetation, McLachlan (1974) found just the opposite: trees formed the poorest of the substrates with the lowest mean faunal weight, which was about a third of that of benthos, and about 18 fold less than that found in aquatic vegetation. The difference could be due to the presence of predominantly hard wood trees with little bark in Lake Kariba, while the submersed trees of the tropical forest and savanna in Volta Lake were rich in thick bark and many trees had soft wood, which became rapidly perforated by the burrows of *Povilla.* In Volta Lake *Povilla* formed 90% of the total standing crop of aquatic macroinvertebrates associated with the bark/wood of submersed trees, followed by the chironomid *Polypedilum* larvae (Petr, 1970). One may conclude that in Volta Lake the massive development of *Povilla* on inundated trees has contributed to maintaining high fish landings from this lake.

Even prior to understanding the significance of the presence of dead trees in reservoirs for fish production through food chain cascade, in some parts of Africa the fish production in shallow water bodies was artificially increased by establishing brushparks. These are made of brush cut around the lake or a lagoon, which, when submersed, will provide surfaces for periphyton and its associated aquatic invertebrate fauna, as well as functioning as fish-attracting devices. The method, first described in Benin, West Africa, under the name acadja (Welcomme, 1972), is now widespread. Annual fish yields vary from 2 t ha^{-1} with fishing intervals of 3 to 4 days, to up to 17 t ha^{-1} with an interval of 70 days (Welcomme and Kapetsky, 1981). The requirement of wood for acadjas is about 10 t yr^{-1} and this may be detrimental to the shoreline brushes and trees. As a widespread use of acadja leads to rapid deforestation around the water body, studies have been made of trees with a reasonable resistance to underwater decay, which could be grown in plantations. The use of brushparks is also common in the Niger River, in some countries of Asia, for example in Indonesia, in the Mekong River, and in Ecuador (Welcomme, 1985).

Function of Drawdown, Macrophytes and Fish Production

In tropical and subtropical countries, the highest water level in a reservoir is normally achieved at the end of floods. The drawdown period follows, during which areas of various extent are exposed. The magnitude differs in different lakes, but may be substantial, as for example in reservoirs used for irrigation purposes, or some reservoirs constructed for hydropower production. Depending on the type of soil exposed and the amount of precipitation available during the exposure, the drawdown area develops terrestrial vegetation. McLachlan (1971) and McLachlan (1974) based on their observations from Lake Kariba, a reservoir in Africa, noted that drawdown functions as part of the lake/ reservoir edge nutrient cycle. Plants which grow on the exposed littoral may be grazed by wildlife or domestic animals, which enrich the drawdown area with manure. The magnitude of drawdown in man-made lakes may be substantial.

After the reservoir Lake Volta filled up completely for the first time in 1968, the drawdown in 1969 exposed approximately 10% of the total reservoir area, this being approximately 80 000 ha (Petr, 1975). During the exposure many plants become established and grow to maturity, to be flooded by the rising water level during the next flood season. When reflooded, their surfaces provide substrate for periphyton and numerous aquatic invertebrates, and the subsequent decomposition of the plants results in high nutrient levels in the shallow water. Water above the reflooded land, enriched by dung of terrestrial mammals grazing on plants during the drawdown period, shows an increase in water conductivity, alkalinity and potassium. This leads to local increases in primary and secondary production.

The regular drawdown and reflooding is equivalent to 'flood pulse' on natural floodplains, which is responsible for their high productivity. In Volta Lake this reservoir flood pulse has probably played an important role in the high fish production of this reservoir. In Lake Naivasha reflooding of the drawdown led to an explosive development of ostracods, and the same was observed in experiments when containers of young papyrus transplanted directly from the lake were flooded (Gaudet, 1977). The rapid build-up of ostracods is accompanied by death and decay of the drawdown flora, increases in other invertebrates, appearance of juvenile fish, increase in fishing bird activity etc. Offshore primary production also increases as the water level rises. In short, the reflooding phase coincides with a period of increased production in a water body.

In Hong Kong, China, analysis of five species of fish (*Carassius auratus, Hemiculter bleekeri, Culter erythropterus, Cirrhina molitorella, Oreochromis mossambicus*) showed that after a rise in water level in Plover Cove reservoir the flooded terrestrial vegetation (sedges, Polygonaceae, and grasses such as *Echinochloa* sp.) became their major dietary items (Dudgeon, 1983). In *C. molitorella* and *O. mossambicus* the terrestrial macrophytes were the dominant component of their stomach-gut contents. The information collected strongly suggested that fish migrated into the marginal zone to feed upon the flooded material.

Management approach, using replanting and restocking, has been applied on the oligotrophic Paizhong Lake on the lowlands of the Yangtze River. This lake originally had a dense growth of macrophytes. After overstocking it with grass carp in 1975, the macrophytes virtually disappeared by 1979. By the 1980s, the grass carp production was negligible, and it was therefore decided to rehabilitate the aquatic plants by sowing the littoral zone with *Echinochloa crus-galli* (213 ha) and *Zizania latifolia* (5.5 ha). Within two months the vegetation biomass reached 10 tons ha^{-1}. The plants were inundated by rising the water level, and subsequently fish were stocked, including grass carp, bottom-feeding and planktivorous species. Herbivorous fish were stocked first, to utilize the aquatic plants and to remineralize the nutrient from aquatic plants into the environment. Bottom-feeding fish were stocked to feed on detritus from the plants and, finally, planktivores were added. Under this system, the production reached was 460 kg ha^{-1}, a 62% increase in production on years when there was no vegetation in the marginal zone. This approach is now widely adopted in China lakes (Zhang and Chang, 1994).

FISH DENSITIES, STANDING CROP AND PRODUCTION, AND AQUATIC MACROPHYTES

Aquatic macrophytes can contribute to an increase in fish abundance when compared with areas or water bodies devoid of macrophytes.

Borawa *et al.* (1979) found that in the Currituck Sound (USA) fish density increased from approximately 1 000 to more than 15 000 fish ha^{-1} after *Myriophylum spicatum* became established. Killgore *et al.* (1989) in their study of fish in the Potomac River (Virginia, USA) found densities of 17 000–98 000 fish ha^{-1} in areas with plants and the CPUE was two to seven times higher in areas with plants than without plants. Killgore *et al.* (op. cit.) pointed out that estimates of fish density in aquatic plants are inherently variable due to sampling techniques, plant densities and patchy fish distributions. Seasonal changes in density and species composition of aquatic plants cause a transition in the spatial and temporal distribution of fish. In temperate and cool temperate zones, during the spring fish tend to associate for food and cover with aquatic plants that overwinter or emerge early from the substrate.

The fish disperse to the more established macrophyte beds in the summer and autumn (Hall and Werner, 1977). Once the plant community reaches its maximum density in the summer, species capable of establishing dense stands, such as hydrilla, can occupy the entire water column, decreasing fish movement and foraging efficiency. Therefore, fish abundance and condition are often higher in areas of intermediate levels of structural complexity, particularly for piscivorous species (for references see Killgore *et al.*, 1989). In aquatic plants in the Potomac River piscivorous fish were commonly observed occupying small "holes" (spaces) devoid of plants within dense vegetation, from where they ambushed their prey. Too dense concentrations of aquatic plants may lead to a reduction in the condition of fish, resulting in a stunted fish population (Cole and Shireman, 1980). In a study on Great Lakes (Lake Huron and Lake Ontario, Canada) the fish production index, based on fish abundance and fish size, was significantly higher at littoral sites with macrophyte beds than at sites where macrophytes were sparse or absent (Randall *et al.*, 1996). Fish size was consistently smaller at the high density macrophyte sites, which supported the findings of Barnett and Schneider (1974) who observed that as plant density increased, greater numbers of smaller fish were found.

Vegetated Lakes

Summarizing information from the literature, Hoyer and Canfield (1996) listed the most important environmental factors affecting the abundance of aquatic macrophytes in lakes as: general water chemistry, lake trophic characteristics, substrate characteristics, light availability, prevailing winds and lake morphology. These factors can work independently and in combination, varying with the scale of analysis. The factors also suggest that lakes can be divided into four general groups, small and large shallow lakes with abundant aquatic macrophytes, and small and large deep lakes with sparse aquatic macrophytes. The littoral zone of lakes is inversely related to basin slope, depth, and to the shoreline development. When lakes are shallow and other factors are favourable, aquatic macrophytes may cover a large portion of the lake. Studies have shown significant relations between aquatic macrophyte abundance and lake water chemistry; phytoplankton population structure and biomass levels; sediment resuspension and wave action; periphyton and aquatic invertebrate populations, and many other limnological processes; fish growth, abundance and population structure; angler utilization of fish populations; aquatic bird abundance and species composition, etc. The magnitudes of these relations are generally in proportion to the abundance of aquatic macrophytes.

In deep lakes, which have less littoral area for aquatic macrophyte growth than shallow lakes, the importance of aquatic macrophytes to the overall functioning of these lakes decreases proportionately as lakes get larger and deeper (e.g. Tilzer and Serruya, 1990). In some cases, small areas of littoral habitats may play a limiting factor for the reproduction or recruitment of some aquatic organisms in large lakes but not the overall production of the organisms.

Based on the investigations of 60 Florida lakes Hoyer and Canfield (1993) came to a conclusion that total and harvestable fish biomass per unit of adjusted chlorophyll *a* is at a maximum when the percent volume with aquatic macrophytes ranges from 20% to 40%. However, some of their lakes had high fish biomass also at macrophyte covers of less than 15% or over 85% thus showing that only a potential for depressed fish populations exists at both low and high levels of aquatic macrophytes.

Bettoli *et al.* (1992) found a strong relation between the submersed vegetation and largemouth bass in Lake Conroe (8 100 ha) in Texas and in Guntersville reservoir (28 000 ha) in Alabama. In Lake Conroe, the density of age-1 largemouth bass declined from about 100 fish ha^{-}

^{1} when submersed vegetation covered 30–44% of the reservoir to about 20 fish ha^{-1} when vegetation was completely removed. In Guntersville reservoir, young-of-the-year largemouth bass densities averaged 350 fish ha^{-1} in submersed vegetation compared to 24 fish ha^{-1} in unvegetated habitats during two years when flushing rates were high. Maceina (1996) cautions that the results of research on aquatic plant-largemouth bass population interactions conducted in different size water bodies, including reservoirs, may not be transferable. His suspicion was confirmed in a study by Hoyer and Canfield (1996a). When they examined the relationships between abundance of aquatic macrophytes and largemouth bass (*Micropterus salmoides*) for 56 lakes in Florida they found that lakes with higher nutrient concentrations tended to have more adult largemouth bass, but there was no apparent relationship between adult largemouth bass stock and abundance of aquatic macrophytes. They pointed out that lakes with grass carp, virtually free of aquatic macrophytes for ten or more years, supported the same numbers of largemouth bass as lakes with macrophytes.

The authors stressed that any study attempting to elucidate relationships between the abundance of aquatic macrophytes and the abundance of largemouth bass, an important sport fish of USA waters, must consider the effect of lake trophic status. From the management point of view, the recommended 20–30% coverage of submersed vegetation (as recommended for example in Texas and Florida) is not needed for largemouth bass. One factor that may contribute to largemouth bass surviving in lakes of different trophic status and in lakes with and without aquatic vegetation is the opportunistic feeding behaviour of largemouth bass. As aquatic macrophytes are removed from the system the bass food habits shift from macrophyte-oriented species (e.g. small centrarchids) to open-water oriented species (e.g. clupeids). However, the authors also pointed out that large lakes may need aquatic macrophytes as refuge for young-of-the-year and subadult largemouth bass more so than small lakes due to the reduced shoreline to surface area ratio.

Reservoirs with Aquatic Plants

In a temperate climate, new reservoirs are not always hospitable environments for establishment of submersed plants. Because many reservoirs are operated for flood control, water supply, irrigation or hydroelectric power production, their water level may fluctuate considerably. Flooded terrestrial soils also may not be suitable for rooted aquatic plants, and wave action along the shoreline may soon wash out fine soil, leaving only pebbles or rocks. While stocking reservoirs with fish has been practised for some time, planting desirable aquatic plants as an important component of the ecosystem on which fish and other organisms depend, remains still only an idea.

Smart *et al.* (1996) believe that more experimental work using a variety of locations and species would contribute to developing methodologies for establishing aquatic macrophytes in many reservoirs still lacking them. Planting of submersed macrophytes has been successfully applied in some lakes in China, such as Hongze, with the purpose of enhancing the production of mitten crab (*Eriocheir sinensis*) (Petr, unpublished), and in the oligotrophic lake Paizhong on the Yangtse lowlands to increase fish production. In Lake Puckaway, Wisconsin (USA) aquatic macrophytes were replanted after a biomanipulation removal of weed fish (Brege, 1986).

Reservoirs constructed on the Volga, Dnieper and Dniester rivers in Russia and Ukraine are large shallow water bodies, which serve in the first instance for hydropower production, some also as a storage of water for irrigation. Four reservoirs were constructed on the upper Volga between 1937 and 1957, the largest being the Rybinsk. Fish stocks have been dominated by the phytophilic bream (*Abramis brama*) and roach (*Rutilus rutilus*) (long-term average in landings 41% and 19%, respectively), followed by pike (*Esox lucius*) (14%), and pikeperch (*Stizostedion volgensis*) (13%). Poddubny and Galat (1994, unpublished manuscript) commented on the decline of aquatic macrophytes in the upper Volga reservoirs, which has resulted in declining catches. While approximately 28% of the Rybinsk are shallows, only 2% of them have aquatic macrophytes, and a constant wave action is attacking the remaining submersed and emergent plants.

Vdovenko and Kravchenko (1980) attempted to determine the optimal density of aquatic macrophytes for bream (*Abramis brama*)-dominated small (200 to 400 ha) shallow reservoirs situated in the lowlands of the lower Don River. With an increase in density of emergent aquatic macrophytes (*Typha angustifolia, Scirpus lacustris* and *S.*

maritimus) there was also an increase in the density of zooplankon - until a certain level: in dense stands of macrophytes zooplankton density was low, probably because of the low concentration of phytoplankton and low dissolved oxygen concentration. Bream was found to breed in areas with a very low density of macrophytes, and this notwithstanding that in denser stands there was a higher zooplankton density. The reason for this could be the deterioration in water quality towards the beginning of June, when concentration of dissolved oxygen starts declining, and there is an increase in concentrations of carbon dioxide and humic acids. Vdovenko and Kravchenko (1980) suggested that thin stands of emergent aquatic macrophytes should cover 30–40% of the productive zone to improve bream production in the investigated reservoirs.

In large lowland Ukrainian reservoirs, where because of their size, water level fluctuation is minimal and very gradual, aquatic macrophytes are well developed. In these reservoirs phytophilic fish represented 85 to 90% of the total fish biomass (Taran, 1990), much of it produced in shallows down to 2.5 m depth. These shallows account for 8–10% of the total area of reservoirs, with the reservoirs subjected only to a small summer drawdown of up to 1.5 m. The shallows are characteristic for a well-developed belt of macrophytes, which harbour rich zooplankton and bottom fauna. The production of organic matter in this zone is estimated to be 2.6 higher than that in the open waters.

The littoral of Dnieper reservoirs is responsible for 85–90% of the total reservoir fish production. The littoral is gradually overgrowing with emergent vegetation, which causes a decline in fish production and waterfowl, and decline in zooplankton production on which many fish feed. Zimbalevskaya *et al.* (1984) believe that aquatic macrophytes should not exceed 50% of the total area of the littoral zone. A gradual degradation through accumulation of organic detritus in the littoral, leading to peat formation and gradual conversion of the littoral plant zone into a swamp, has been taking place in a number of the Dnieper lowland reservoirs.

Land improvement to enhance spawning opportunities for phytophilic fish has been tried on the lower Don River. Tsimlyansk reservoir belongs to the most productive reservoirs in the Ukraine, with a 20-year annual average catch of 10 700 t of fish (Nefedov and Fesenko, 1987), and a yield ranging between 35 and 58 kg y^{-1}. To maintain the high yield, deterioration of floodplain lakes had to be halted. The lakes represent an important spawning habitat, but they have been gradually overgrowing with wetland plants. Part of the vegetation was therefore removed reducing the cover from 20–30% to 5–10%.

This resulted in a five-fold increase in spawning rate as compared with the unimproved floodplain lakes. Access to floodplain lakes was also improved by deepening the channels connecting them with the reservoir. As with other hydropower producing reservoirs, Tsimlyansk is subject to a drawdown, which may expose up to 40 000 ha of land. When fully reflooded in late spring, this area provides the best spawning grounds on flooded terrestrial grasses. The authors make a point that maintaining passable channels for broodstock to reach floodplain lakes, and prevention of floodplain lakes from getting overgrown with emergent macrophytes, is not cheap.

Exact data on fish yields from vegetated areas of tropical reservoirs are difficult to come by. In agreement with the observations on temperate and subtropical water bodies, the presence of aquatic macrophytes in the littoral of tropical reservoirs also increase fish production. In the Tanzanian reservoir Nyumba ya Mungu, where a significant portion of the littoral zone was occupied by aquatic macrophytes, the presence of the more open-community emergents *Paspalidium germinatum* and *Cyperus articulatus* determined the presence of major shoals of the commercial fish *Sarotherodon pangani* and *S. jipe*, which feed largely on periphytic algae growing on aquatic plants, branches of inundated trees, and stone surfaces (Bailey *et al.*, 1978).

In an experimental fishery, these two species together constituted 72% of the total (Bowker and Denny, 1978), while *Tilapia rendalli*, species introduced into the upper catchment of the reservoir, and the only fish specialized in grazing aquatic macrophytes, constituted 7%. The rapid growth of *Paspalidium* and *Cyperus* takes place in the nutrient enriched waters, and the same nutrient enrichment is also responsible for a massive growth of periphyton on which *Sarotherodon* feed.

Floodplains

Floodplains represent areas where the rise in water during the floods inundates extensive areas of land flanking the river channel. Frequently, such a rise in water level connects the river with floodplain lakes, isolated from it during the dry season. While flooding dry land results in covering largely terrestrial grasses and soil with water, reconnecting the river with floodplain lakes provides the fish of the river channel with the opportunity to access their areas covered with aquatic macrophytes, and where such lakes were under a high fishing pressure during the low water level, to repopulate such lakes with fish.

The increase in living space, together with the release of nutrients associated with the submersion of the soil and from the decaying plants produces an annual surge of plankton and periphyton, encouraging an explosion in the animal productivity. The present publication does not deal with the significance of seasonal changes on floodplains, and the reader is referred to other publications, such as Welcomme (1979, 1985, 1989) and Gaudet (1992), who deal in more detail with the floodplains of Africa, Howard-Williams and Gaudet (1985) and Thompson (1976) who describe the structure and functioning of African swamps, and to Junk and Weber (1996), Junk (1996) and Welcomme (1979, 1985) who discuss the Amazonian floodplains. Welcomme (1985) constructed a diagram of trophic relationships in a river-floodplain community.

As an indication of the importance of aquatic macrophytes for the productivity of floodplains in the Amazon varzea may serve the following estimate: the floating meadows of *Echinochloa polystachya* can accumulate up to 80 t ha^{-1} biomass, much of which is deposited on the exposed drawdown area after the retreat of flood waters. The decomposition of this plant biomass, together with that of the terrestrial plant species *Paspalum fasciolatum* plus other herbaceous plants, which grow on the exposed soils during the dry phase, plus from leaf litter etc. from the inundated forest, maintains the high productivity of the floodplain and its water when flooded, and this is reflected in the high fish production. The flood pulse of the Amazon and many other rivers, but especially those with substantial floodplains, is recognized as a major factor in maintaining the high productivity of floodplains.

Rivers

Submersed aquatic plants that are abundant in some stream reaches have a potential to provide winter concealment cover for fish. Juveniles of several salmonid species were found in winter in rooted aquatic vegetation in a spring-fed Credit River, Ontario, Canada, in portions of the Kenai River, Alaska, and in a Norwegian river (Griffith and Smith, 1995 and references therein). However, in the Henry's Fork of the Snake River in Idaho, during two years of observations, the cover provided by submersed macrophytes was not adequate to hold 0+ rainbow trout throughout the winter. During one winter the loss of macrophytes resulted from sloughing and foraging by swans and other waterfowl, during another winter the loss was due to the presence of anchor ice (Griffith and Smith, 1995).

Hoyer and Canfield (1992), taking total fish standing crop data from North American streams located in Wyoming, Vermont, Florida, Iowa, Ontario, Washington and Missouri, noted that the regional estimates of average fish standing crop were significantly correlated to the total phosphorus concentration. Total fish standing crop increased with total phosphorus concentrations throughout the range of reported values, but increased much more rapidly in streams having phosphorus concentrations e" 15 ì L^{-1}. However, the results of a survey of 17 Florida streams have shown that macrophyte standing crops were not correlated with instream total phosphorus or total nitrogen concentrations. This suggests that regional differences will be masked when included in a larger geographical coverage.

Mean fish density within the stands of aquatic plants of the Potomac River (USA) ranged from 5 000 to 204 000 fish ha^{-1} (Killgore *et al.*, 1991). The establishment of aquatic plants had a positive influence on the fish abundance, particularly juvenile sport fish (largemouth bass, yellow perch, pumpkinseed): over 300 largemouth bass were collected in areas with aquatic plants, while only one individual of this species was collected in areas without plants. Considering all fish species in total, up to seven times more fish were collected in areas with plants.

Referring to literature, Killgore *et al.* (1991) provided comparative figures of fish densities in aquatic plants for some other water bodies: in Currituck Sound fish density increased from 1 000 to more than 15 000 ha $^{-1}$ after Eurasian watermilfoil (*Myriophyllum spicatum*) became established. In Orange Lake, Florida, fish density ranged from 13 000–205 000 fish ha $^{-1}$ in areas with submersed aquatic plants, and up to 86 000–2 500 000 fish ha $^{-1}$ in several lentic locations in central Florida.

Managing in-stream and riparian vegetation for the enhancement of fish stocks or to provide better conditions for fish species preferred for fishermen, has been undertaken for centuries, mainly in the temperate zone of the northern hemisphere. Because of the vital links between river vegetation and water quality, physical habitats and animals, the manner in which vegetation management is addressed is critical for fishery (Cowx and Welcomme, 1998). The authors have pointed out that without management and the great importance attached to some highly-priced lowland fisheries, many rivers would have been virtually destroyed in the headlong drive for floodplain drainage. See Cowx and Welcomme (1998) for more details on the control of vegetation in water courses, on establishing aquatic and riparian vegetation to benefit fisheries, and on utilizing vegetation to enhance fisheries.

Canals

In the 1980s there were over 200 million ha of land under irrigation in the world, and the increase was estimated to be 20% over 10 years (Van Zon *et al.*, 1982). Many irrigation and drainage canals experience a persistent problem of overgrowing with aquatic macrophytes. Spring-fed canals in Europe and Asia, especially in their upper reaches, are characterised by considerable growth of emergent (*Phragmites, Typha*) and other macrophytes including *Bulboschoenus, Schoenoplectus, Potamogeton, Myriophyllum*. Filamentous algae *Cladophora, Spirogyra* and *Oedogonium* are also common.

In most irrigation systems some form of fishery is caried out, but always extensively. The yield of small-scale fishery is estimated to be 10 to 60 kg ha^{-1} y^{-1} (Van Zon *et al.*, 1982). These authors suggested converting aquatic plants, infesting irrigation and drainage canals, to fish protein. This has been tested in Egypt by using grass carp. Sustained harvesting of aquatic plants could produce 200 kg of grass carp per ha, and probably also approximately 100 kg of indigenous fish species.

Under a well managed polyculture system, perhaps 500 kg ha $^{-1}$y^{-1} could be achieved. In the late 1950s the Karakum Canal in Central Asia developed considerable coverage of submersed *Myriophyllum*, which in the early 1960s was successfully controlled by the introduction of grass carp. *Myriophyllum* was succeeded by *Ranunculus*, which is avoided by herbivorous fish such as grass carp. Charyev (1984) summarized the experience with the introduction of grass, silver (*Hypophthalmichthys molitrix*) and bighead (*Aristichthys nobilis*) carps into the Karakum Canal and emphasized the need to protect the higher aquatic vegetation, particularly in cases where macrophytes represent spawning substrate for other fish, such as the common carp.

Common carp was efficiently used for suppressing aquatic macrophytes in drainage canals in Argentina (Sidorkewicj *et al.*, 1998). In one instance 1 000 and 2 000 fish ha^{-1} (average weight 20 g) were applied following the mechanical removal of vegetation; another time 500 and 1 000 fish ha^{-1} (average weight 260 g) were stocked in the existing vegetation. Three months after the introduction of the small fish the regrowth of the dominant submersed plants *Chara contraria* and *Ruppia maritima* was well under control. The larger fish completely destroyed all submersed plants within four months. The activity of the common carp resulted in an increase in turbidity of up to ten times that on control plots. The authors noticed that the common carp also fed on aquatic macrophytes.

AQUATIC MACROPHYTES, RECREATIONAL FISHERIES, AND RECREATION IMPACT

Studies conducted in Ireland and elsewhere demonstrate that water bodies rich in vegetation generally support productive recreational fisheries. This reflects the cover, direct and indirect food supply and spawning substrates that aquatic macrophytes provide for fish and invertebrates. The composition and structure of the vegetation has also been shown to influence fish stock levels, through its impact on macroinvertebrates and periphyton supply, and by interfering with freedom of movement of the fish.

For recreational canals in Ireland with 20% to 70% plant cover, Caffrey (1993) reported the mean fish standing crop of approximately

2.5 times higher than in canals with vegetation exceeding 70%, and 4 times higher than in those with less than 20% cover. Sparsely vegetated sections supported 30–82 kg ha^{-1} of fish. In densely vegetated sections, dominated by *Vaucheria* sp. (filamentous alga), macroalga *Chara* spp, and *Potamogeton pectinatus*, 10 out of 17 sections had less than 100 kg ha^{-1} and represented poor recreational fishery. Most supported mixed fish communities, dominated by bream, although pike and perch, both piscivorous species, were well presented. In moderately weeded canal sections, the lowest fish standing crop was 116 kg ha^{-1}, with 16 out of 24 sections having over 200 kg, and six of these in excess of 400 kg ha^{-1}. The highest mean standing crop of 451 kg ha^{-1} was recorded from the sections where submersed plants with broad leaves, such as *Potamogeton*, or with complex leave types, e.g. *Myriophyllum* spp, charophytes and *Ceratophyllum demersum* dominated. Where strap-shaped or streamlined plants (*Sparganium emersum*, *Scirpus lacustris*, *Sagittaria sagittifolia*) dominated, the mean standing crop was 120.4 kg ha^{-1}. In sections with a mixture of growth forms the mean standing crop was 201.5 kg ha^{-1}, and where algae dominated, the mean standing crop was 148 kg ha^{-1} (Caffrey, 1993).

Good angling may be expected in moderately vegetated canals, which support more than 200 kg ha^{-1} of coarse fish. Denser vegetation stands restrict free movement of open water fish, such as roach, rudd, pike and perch (Canfield and Hoyer, 1992).

In Lake Conroe, Texas, elimination of aquatic vegetation from an area of over 3 600 ha using grass carp led to a change in the structure of the sport fish community (Bettoli *et al.*, 1993) The largemouth bass-crappie-hybrid striped bass fishery was replaced by a channel catfish-white bass-hybrid striped bass-largemouth bass-black crappie community. Whether sport fishing on this lake got better or worse the authors were unable to say. Creel survey results indicated a decline in catch rates and an increase in average size of largemouth bass after vegetation removal. Lower catch rates were probably due, in part, to the lower density of largemouth bass after vegetation removal.

The Lake Conroe experience has several management implications, summarized by Maceina *et al.* (1991). In lakes and reservoirs where crappies (*Pomoxis* spp) depend on threadfin shad (*Dorosoma petenense*) for food, managers may consider control of excessive aquatic macrophytes to enhance crappie fisheries when open-water limnological changes have occurred. Macrophyte reduction should influence the trophic structure of the limnetic fish population, and crappie growth rates should improve. However, Maceina *et al.* (1991) caution that an optimal level of plant coverage for maximizing sport-fish standing crop may not be applicable, because of a number of variables, such as regional geology, water level fluctuation, plant community type, hydraulic flushing rate, basin morphometry, and external and internal nutrient loading rates. There are also differences in the type of fishery to be managed.

In Red Haw Lake, Iowa, reduction in vegetation biomass by 91% in four years using grass carp increased the opportunity for anglers to fish from shore (Mitzner, 1978). During the investigation popularity of shoreline fishing increased by 241% with catch success remaining greater than 0.70 fish per hour. Boat anglers also became more numerous and usually more successful than shore fishermen, and they caught nearly twice as many fish. Bailey (1978), who assessed the impact of grass carp control of aquatic macrophytes in over 100 lakes in Arkansas, expressed the opinion that submersed aquatic weed control with grass carp is far superior to chemical control, mechanical control, or no control. As lakes in the USA are managed for the benefit of fishery, with the ultimate goal to satisfy the greatest number of fishermen, Bailey concluded that the use of grass carp for submersed weed control is an effective and desirable tool.

In New Zealand, temporary removal of aquatic plants in Lake Parkinson, a small, eutrophic lake, resulted in a number of changes to the population of stocked rainbow trout (*Oncorhynchus mykiss*). In the first summer following weed removal the number of cormorants (*Phalacrocorax carbo*) counted at the lake increased and their predation resulted in a decline in trout density. However, the growth rate and condition of the trout population then exceeded that of trout present before weed removal. During the second summer after weed removal a cladoceran bloom was followed by low phytoplankton levels and high ammonia concentrations. These low oxygen levels eliminated the trout population, but other fish species present survived. The experiments demonstrated the importance of weed beds in maintaining a stable fish community in lakes such as Lake Parkinson (Rowe, 1984). In the lower Potomac River, Virginia (USA), where anglers target especially the piscivorous largemouth bass (*Micropterus salmoides*) and yellow perch (*Perca flavescens*), spaces ('holes') within the dense stands of hydrilla were found to provide a favourable habitat for the piscivorous fish. Expansion of such holes, or creation of new ones in dense aquatic plants would also result in better fishing success for such sport fish. Another management method is creation of open lanes (channels) through the aquatic plant beds. Crowder and Cooper (1979) recommended creation of rectangular strips of intermediate vegetation density to create sufficient "edge" for larger fish to forage on small fish. This type of control strategy can allow access by boats, while maintaining a productive interaction between the aquatic plants and fish. Such management approach is more suitable for lakes than rivers.

In the USA, where largemouth bass (*Micropterus salmoides*) is an important sport fish, Trebitz *et al.* (1997) developed a model for heavily vegetated temperate lakes that simulates the interactions of bluegill (*Lepomis macrochirus*), largemouth bass and their invertebrate prey in the context of vegetation structure. Bass growth rates and numbers increased after most simulated vegetation removals because their access to prey increased. Bluegill grew fastest when about 30% of the vegetation was cut, but responded negatively to mowing more than about half the plants. Bluegill responded to the manipulation more by changing growth rates, while bass responded more by increasing numbers. Moderate plant removals accomplished by mowing many narrow channels are most likely to simultaneously benefit both species.

Recreation along river banks and shores may cause a major damage to the littoral aquatic vegetation and through this impact the spawning and nursery habitat of some fish. Along some lakes and reservoirs in Germany and Hungary especially, the summer invasions of tourists seeking water may cause considerable damage to the water plants through trampling, as well as by using boats. Further damage may be caused to terrestrial bank or shore vegetation, the shade of which is known to attract some fish species to the littoral. Pleasure boats may increase water turbidity, which is the result of stirring bottom sediments. This may lead to recycling of nutrients and cause increase in phytoplankton.

Strong correlation has been found in the English canals between the intensity of boat traffic and the decrease in submersed and floating-leaved macrophytes, especially *Potamogeton natans* and *Nuphar lutea* (e.g. Murphy and Eaton, 1981). On the other hand boat traffic performs the useful function of maintaining a satisfactory amount of vegetation against the natural tendency for plant growth to become excessive. The study of 13 canals in England has shown that canals with little or no boat traffic had mostly good fish stocks.

FISH CAPTURE METHODS IN VEGETATED WATER BODIES

Fish density estimates in aquatic plants are inherently variable due to sampling techniques, plant density, and patchy fish distribution. Dibble *et al.* (1996) made a frequency list of the methods and tools used to quantify fish in or near vegetated areas. The numbers correspond to individual papers listing a particular method: electro-shocker (11); divers (11); seine (10); rotenone (8); popnets (7); gillnets (7); drop or throw nets (6); light trap or minnow trap (6); explosives (6); belt transect (5); rotenone and block net (5); modified traps (3); push net (3); underwater camera (3); electro-shocker and block net (3); modified nets (2); strip counts (2); ten other methods, each used once.

Popnets, drop nets and throw nets seem to be most effective in sampling fish in aquatic macrophytes. Popnets and drop nets measure distribution, diversity and abundance of adult and juvenile fish in densely vegetated areas where traditional methods (i.e. seining and electrofishing gear) are ineffective. Killgore *et al.* (1988, 1991) and Morgan II *et al.* (1988) used popnets in their study of fish in *Myriophyllum* in Lake Guntersville in Alabama, and in *Hydrilla* beds in the Potomac River (USA), respectively. Variability between individual popnet collections were high for each site in the Potomac River indicating patchy distributions within the *Hydrilla* beds. In contrast to techniques that require the collection of stunned (electrofishing) or dead (rotenone) fish with a dip net in dense plant beds, the popnets reduce the potential bias of underestimating fish abundance that is associated with the above methods. Larson *et al.* (1986) reported that the sampling efficiency of the popnets was near 100 percent. Freeman *et al.* (1984) who used a drop trap (1 m^2) recovered 90 percent of tagged fish. Enclosure traps, such as drop nets and popnets, may have the least source of bias in estimating the density of fish residing in dense plant beds because all fish within the net can be collected or accurately estimated (Killgore *et al.*, 1988). This technique can also be easily replicated. However, fish biomass estimates may be underestimated because larger fish can escape more easily than smaller fish. Killgore *et al.* (1988) also commented on the use of electrofishing in plant beds: the efficiency of electrofishing generally decreases as plant density increases because of the difficulty in locating and collecting stunned fish in dense vegetation.

Chick and McIvor (1994) in their study of fish in macrophytes of Lake Okeechobee used for sampling the fish a circular drop trap and square throw trap. Both of them were found to be effective for sampling fish in shallow vegetated habitats.

In two freshwater streams in Costa Rica to obtain estimates for fish biomass per unit area with floating aquatic plants (*Eichhornia, Hydrocotyl, Salvinia, Azolla*) a small seine net was hauled moving the leadline under the floating vegetation mass, then quickly raising the trapped vegetation and fish above the water and removing them from the net at the shore or in a boat. The area of the haul was the area of the seine net (Winemiller and Ponwith, 1998).

In a study of fish in the varzea lake Mamiraua, situated between the rivers Solimoes and Japura, the fish within floating vegetation mats were sampled quantitatively using a 23 m sand eel seine net. The net was made up of five panels of differing mesh size. The vegetation was sampled by encircling an area of 2 to 12 m^2 with the net deployed from a boat. The lead line was then pulled under the root mat and the mass of plants pulled towards the boat (Henderson and Hamilton, 1995).

When rotenone sampling in vegetated areas the efficiency of fish recovery decreases with the increasing plant density (Shireman *et al.*, 1981). Seines were commonly used near vegetation, but were difficult to use in dense plant beds.

Electrofishing has been used in numerous studies. However, electrofishing is known to be selective, with smaller fish being less vulnerable to electric shock than larger fish (Zalewski and Cowx, 1990). The presence of vegetation may also hide fish fleeing from the electrical current, and contribute to a lower catchability (Randall *et al.*, 1996). Rider *et al.* (1994) compared rotenone and electrofishing for largemouth bass in vegetated (*Myriophyllum spicatum*) and non-vegetated areas in Guntersville Reservoir in Alabama. The results from both methods used were not significantly different, but using rotenone was more costly than electrofishing.

Using grass carp in aquatic macrophyte control raises sometimes a problem of its recapture, after the job has been done. In a small water body of only 0.2 ha in Florida, USA, Mallison *et al.* (1994) used rotenone bait, which succeeded in removing up to 79% of the triploid grass carp present. Using Fish Management Bait (FMB), as the authors call it, may be feasible in lakes if moving feeders consistently attract naive fish and if the majority of the triploid grass carp population can be attracted to feeders. FMB was selective to triploid grass carp and resulted in minimal kill of the native fish species. Public angling has also been used as a method of triploid grass carp removal (Mallison *et al.*, 1994a). Trammel and gill nets were successful in recapturing grass carp in Luisiana (Richardson, 1994). Bonar *et al.* (1993) evaluated the effectiveness of various capture methods of grass carp in a vegetated lake. Herding of fish was the most effective of the seven techniques, followed by angling in baited areas. Herding removed 0.8% to 8.2% of the carp stock in one sweep in lakes containing dense vegetation, submersed logs and other underwater obstructions. The authors recommended this method for reducing the numbers of fish in overstocked water bodies smaller than 10 ha, or for capturing fish to monitor growth. Electroshocking has a potential as a herding device.

5

Aquatic Plant Management and Fish

MANAGEMENT IMPLICATIONS

The advantage of aquatic plants for fish species diversity and biomass is well documented from numerous temperate, subtropical and tropical water bodies. An increasing number of detailed field studies is being supported by controlled field and laboratory experiments which allow to better understand the often complicated relationships of water quality, primary algal production, zooplankton, micro- and macrobenthos, periphyton, aquatic macrophytes and fish. Regulation of aquatic macrophytes can be executed through interference with most of the individual links within the web of the interrelationships, such as for example nutrient concentration, periphyton, zooplankton, macroinvertebrates (e.g. crayfish), or the individual species of fish and aquatic macrophyte communities. Some of the knowledge has been applied in the biomanipulation approach to improving water quality in shallow eutrophic lakes of the temperate zone. Manipulation of aquatic plants has led to an increase in desirable sport fish, or to an overall increase in the productivity of a water body. Control of macrophytes using fish has been used in achieving higher fish production, especially in small water bodies, such as ponds and small lakes.

Occurrence in some waters of invasive aquatic macrophytes, such as *Hydrilla* and *Myriophyllum*, although not appreciated by some water users, has often led to a considerable, multifold increase in fish stocks. Floating exotic invasive plants such as *Salvinia molesta* and water hyacinth, if not occurring in a solid mass over large areas, have been found to also enhance fish stocks by providing shelter and feeding opportunities. In tropical waters high fish catches are achieved along the margins of lakes with floating aquatic plants, such as papyrus.

Knowledge of fish behaviour is an important factor in aquatic plant management for fishery purposes. Many fish associate with either the water surface immediately above the plants, the periphery of the dense stands, "holes" formed in the plant beds, or the bottom directly below dense canopy formations to use both open and structurally complex areas for foraging and predator avoidance (Killgore *et al.*, 1989). In dense plants piscivorous fish may be more common at the periphery of the plant beds or in the "holes" in aquatic plants. Creation of strips of intermediate vegetation density or open lanes within the vegetation is thought to enhance the stocks of piscivores and the access of sport fishermen to these fish. Olson *et al.* (1998) removed aquatic plants from about 20% of the littoral zone by cutting a series of channels through each of the four investigated lakes in Wisconsin (USA). In the first year after manipulation, bluegill and largemouth bass substantially increased growth rates of some age classes. The results suggest that harvesting macrophytes in a series of deep channels may be a valuable tool for integrated management of fish and macrophytes.

On the other hand, when aquatic macrophytes occupy large areas of a water body, a reduction in the fish condition can occur, resulting in a stunted fish population (Colle and Shireman, 1980). Annual senescence of the plant community in temperate waters will result in a gradual reduction in refuge for small fish, exposing them to predators. Or, the senescence will result in the loss of epiphytic algae and associated invertebrates, leading to a decline in the condition of fish depending on this food.

Fishery objectives should be addressed prior to any generalised management orientation regarding aquatic vegetation control. Aquatic vegetation represents an important habitat for fish, which may use it for feeding, reproduction and escape from predators. Aquatic macrophytes also play a vital role in bringing nutrients trapped in bottom sediments back up into the aquatic environment, and the detritus generated by aquatic macrophytes may boost the production of benthic organisms. Controlling macrophytes using grass carp enhances fish stocks of some other fish species.

Aquatic Vegetation - an Obstacle to Fishery

A gradual or fast increase in one particular aquatic macrophyte of an invasive character can result in a threat to commercial or recreational fishery. In many lakes throughout the tropics, water hyacinth or salvinia has made commercial fishery difficult. This applies to large lakes of East Africa, reservoirs in Asia and Latin America, floodwater lakes (lagoons) in Papua New Guinea, and many other water bodies. Massive spread of the indigenous floating macrophyte *Pistia stratiotes* on one arm of the Volta Lake (reservoir) in Ghana has also represented an obstacle to fishery activities. Sudd and floating islands in Sudan and some other places of Africa, and varzea meadows in the Amazon Basin also hinder fisheries. In some Florida lakes invasive *Hydrilla verticillata* or *Myriophyllum spicatum* have required a management response to prevent a decline in recreational fishery. Biological control using fish has been successful only in some situations concerning submersed plants, but fish have made little impact on floating plants.

Over the last 20 years Lake Seminole, Alabama (USA) has experienced a dramatic increase in the coverage of *Hydrilla verticillata*, peaking at about 65% surface area in 1992. A comparison of the angling effort and total catch showed a decrease by over 36% between 1985 and 1996, and nearly all of this decline was attributable to fewer anglers seeking largemouth bass (*Micropterus salmoides*). In 1996, catch rates of largemouth bass greater than or equal to 305 mm total length were extremely low. Catch rates for sunfish (*Lepomis* spp) also decreased (Slipke *et al.*, 1998). Interestingly enough, in a Florida lake in the presence of a coverage by 75% of *Hydrilla* the survival of the young pike *Esox niger* and some other American sport fish was higher than outside such areas (Shireman *et al.*, 1983).

In Lake Kyoga, an important fishery water body in Uganda, mats of water hyacinth have blocked the access to numerous fish landing sites on the southern part of the lake. Along the eastern and southern shores a 10 to 15 m wide belt of water hyacinth forms a fringe along the papyrus. In 1995 four ferry crossings were also often completely blocked by this plant (Orach-Meza, 1996). Not only does this plant cause blockages, but it also swamps fish breeding sites and hinders fishing, which has been reflected in reduced catches, rising costs of operation also due to the higher consumption of fuel, lower incomes to the fishermen and higher prices to consumers. The invasion of the bays by the weed has caused fish to migrate to open waters, particularly in Lake Victoria, and has thus reduced the potential of the bays for fishing, spawning and nursing. Along the Njira beach, Kenya, water hyacinth blocked the fish landing sites. To get to the shore, the fishermen had to clear the plant manually.

The plant also interfered with fishing nets set to catch *Rastrienobola argentea*, a small common pelagic fish, which is recorded as second in importance after the Nile perch. This fish, dried in the sun, is sold for animal and poultry feed. To get around the problem of masses of water hyacinth, some Lake Victoria fishermen started installing fish traps around the plant patches, as seen along the Yala swamp in Kenya by Thompson in 1991 (FAO, 1993a). By 1999, in Lake Victoria water hyacinth has been reported to cover already 10 000 ha, and a united strategy for fighting this problem facing the three countries sharing the lake, has yet to be found (Lindsey and Hirt, 1999). The currently dramatic reduction in water hyacinth cover as a result of biocontrol measures suggests (see Section 7), however, that an effective control measure is now available, making the formulation of future control strategies for the riparian countries much easier.

In Papua New Guinea, the explosive growth of *Salvinia molesta* in the late 1970s in the Sepik River lagoon system, was a major obstacle to fishing. Prior to the appearance of this plant, there was a steady growth in catches of *Oreochromis mossambicus* from the lagoons. It was expected that this introduced fish would form a base for a salt and smoked fish production on a relatively large scale. The expansion of *Salvinia* over some 250 km^2 of lagoons led to a collapse of this fishery in the late 1970s. The situation improved in the early 1980s, when *Salvinia* was largely destroyed by the introduced beetle *Cyrtobagous salviniae* (Photo 20).

Further fisheries enhancement has been achieved by the introduction of a second tilapia, *Tilapia rendalli*, which has become well established in the lowland swamps of the Sepik-Ramu river system. In Kerala (India) backwaters, the seasonal explosive growth of *Salvinia* virtually stops any kind of fishing there for several months.

In the Philippines, water hyacinth has caused a major problem to netcage culture production in Laguna de Bay.

Chemical Control of Aquatic Plants and its Impact on Fish

It is not the purpose of this publication to provide a comprehensive review of side-effects of the individual, most common herbicides used today to control emergent, floating or submersed aquatic macrophytes. A few examples in the following paragraphs highlight the impact on fish of the most benign for non-target organisms herbicides glyphosate and 2,4-D (used for floating and emergent plants), diquat and fluridone (used for submersed plants).

Glyphosate, the active ingredient of the herbicide Roundup, has been cleared by the World Health Organisation for safe use in or near watercourses, being virtually non-toxic. Trials in Ireland have demonstrated the capacity of glyphosate to remove obstructive stands of reeds (mainly *Schoenoplectus lacustris, Glyceria maxima, Phragmites australis, Sparganium erectum* and *Typha latifolia*), to create unobstructed areas for anglers in a stretch of a river that was virtually unfishable for years. Trout (*Salmo trutta*) and salmon (*Salmo salar*) used the newly exposed gravel for spawning in the winter, thereby improving fish recruitment and production in the fishery. The cleared areas remained open for three years following a single application (Caffrey, 1996).

Herbicide treatment, resulting in the removal of aquatic plants, also removes associated epiphyton on which some fish feed, and this will lead to a decline in certain fish species. Where aquatic macrophytes are essential for some species for some part of their life history, such as spawning, as a nursery site, the species will also disappear after a treatment with herbicides. It has been observed that in Irish canals application of herbicides (dichlorobenil) reduced the flora and significantly reduced fish habitats, consequently the stocks (Caffrey, 1993).

Glyphosate has been also used for treatment of vegetation in ponds in Poland prior to their flooding (Krueger *et al.*, 1996). This has resulted in the intensive development of food organisms on which carp (*Cyprinus carpio*) feeds. The yield was greater than merely flooding the vegetation or the alternative technique of maintaining a bare fallow prior to flooding. Using glyphosate as part of the pond management programme proved cost effective and had no deleterious effect on the carp fry or their food organisms.

Control of water hyacinth in field trials in the Niger Delta, Nigeria, using glyphosate at 2.88 kg ha^{-1}, resulted in all plants being killed 7 days after treatment. The dead plants sunk or were removed. Subsequent fish catches have shown that the number of some species increased after the treatment, and 28 days after the treatment the catches contained high numbers of juvenile and immature fish (Olaleye and Akinyemiju, 1996). The increase in fish is considered to be the result of improved water quality after the removal of herbicide-treated water hyacinth.

Control of water hyacinth on small areas using 2,4-D is effective, and while fish are not being poisoned by the herbicide, the elimination of the weed in some waters may have an effect on fish which feed among its hanging roots and take shelter there. The herbicide 2,4-D is widely used in fish ponds for controlling aquatic weeds and algal blooms. While some studies suggested a negative impact of this herbicide on survival and growth of fish (Smith, 1952), others have noted a beneficial influence, such as for bottom-feeding fish (e.g. Sarkar, 1991). The rate of application was found to be of importance, with application of 2,4-D at 4.3 kg ha^{-1} being non-toxic for tilapias (Khogali and Moghraley, 1979).

As a response to the treatment of aquatic weeds in fish ponds in India at concentrations of 4.5 to 10.5 kg ha^{-1} there was an increase in the bottom invertebrate fauna. Consequently, the bottom feeding common carp and mrigal (*Cirrhinus mrigala*) showed an increase in the growth rate as compared to the column-feeding fish catla (*Catla catla*) and rohu (*Labeo rohita*). The reduction in the yield of the column-feeding fish which feed on plankton was considered to be due to the destruction of the plankton following herbicide treatment. Also in India, Patnaik (1976) tested the impact of 2,4-D on water hyacinth, *Spirodela polyrhiza, Azolla pinnata* and *Salvinia cucullata* at concentrations up to 3.0 kg ha^{-1}. He found no harmful effects on fish, i.e. *Notopterus notopterus, Barbus ticto, Channa striatus, Cirrhinus mrigala, Labeo rohita, L. bata, Catla catla, Cyprinus carpio, Hypophthalmichthys molitrix.*

The effect of 2,4-D amine on the movement of largemouth bass (*Micropterus salmoides*) was tested in two Florida lakes in the USA, in response to the complaints of anglers that fish flee from or avoid herbicide-treated areas. Area of water hyacinth was sprayed with the legal limit of 2,4-D amine (4.3 kg ha^{-1}), in which four largemouth bass were implanted with internal radiotransmitters, and tracked for six months. Areas of activity before the treatment and after were not significantly different in size.

The results suggest that the surface application of 2,4-D amine directly affects the movement of largemouth bass (Boyer and Cichra, 1994). In response to another criticism of anglers, that fishing success is greatly reduced in herbicide-treated areas and that largemouth bass avoid bait fish captured from herbicide-treated areas, Boyer and Cichra (1994a) conducted trials on the feeding behaviour of largemouth bass. Prey fish golden shiners (*Notemigonus crysoleucas*) and bluegill (*Lepomis macrochirus*), respectively, were held for 90 minutes in concentrations of 1 to 50 mg L^{-1} 2,4-D amine and fed to the bass. In a third trial, largemouth bass were fed crayfish (*Procambarus paeninsulanus*) after the tank level of 2,4-D amine had been raised to 50 mg L^{-1}. In all trials, the bass fed before and after the addition of 2,4-D amine. Feeding behaviour was not affected by the addition of 2,4-D amine to the prey or into the surrounding water even at 50 mg L^{-1}, a level at least 50 to 100 times higher than that found in water after spraying.

Fluridone is a safe, slow-acting herbicide that provides control of rooted aquatic plants such as *Hydrilla, Myriophyllum, Ceratophyllum, Egeria, Potamogeton, Cabomba* without affecting zooplankton, benthic organisms, crayfish, turtles, frogs, water snakes, fish and waterfowl. It becomes toxic to fish only if the normal rate of concentration is exceeded by 76 to 220 times. Fluridone does not affect water quality. It does not accumulate in fish, and aquatic birds continue to feed from the treated areas (Arnold, 1979). Control of susceptible plant species normally lasts from 12–18 months. Being a slow-acting herbicide, fluridone requires 11 months to achieve a 100% control of hydrilla tubers. The fluridone-tolerant plant species are water hyacinth, *Pistia, Chara, Nitella*, a number of emergent plants and planktonic algae (Tarver, 1986). At recommended field application rates, herbicides likely to be used in conjunction with grass carp are not acutely toxic to the fish (Tooby *et al.*, 1980, Kracko, 1989). However, grass carp may stop feeding due to changes in the vegetation which makes it unpalatable. Interference with photosynthesis or destruction of chlorophyll in a plant will affect carbohydrate production, and may lower the food quality of the plants (McCowen *et al.*, 1979). In experiments, after treating hydrilla with either diquat or fluridone, the fish consumed significantly less treated vegetation (Kracko and Noble, 1993). In water, detectable levels of diquat were found 28 days after treatment, and in sediments detectable levels of fluridone were found up to 86 weeks from the date of treatment. However, where the water after the treatment can be flushed out, as in irrigation canals and drainage ditches, fish are successful in hydrilla control (Canfield, 1983).

In the USA, herbicides have provided the most environmentally compatible, cost-effective means of water hyacinth control. However, this has not been achieved without resistance from anglers and environmentalists, who in July 1986 stopped the use of chemical control in Lake Okeechobee, the largest of the Florida lakes. A moratorium on the use of herbicides and instruction to use mechanical harvesters had considerable negative consequences: within 6 months the area of water hyacinths expanded from 810 hectares to 3 240 ha, obstructing most of the boat trails and canals, as well as the 22 flood control and navigation structures. The herbicide programme was resumed immediately to restore order but not before thousands of acres of native vegetation, beneficial to fisheries and waterfowl, were smothered or uprooted by floating mats of water hyacinth (Schardt, 1999)

At present, from 450 public water bodies covering more than 500 000 ha of fresh water, about 300 contain water hyacinth. This plant once covered 50 000 ha of public waterways, but since the middle 1980s it has covered less than 2000 hectares at any one time. About 10 000 ha of water hyacinth are annually managed at a cost of about US$ 2.5 million to sustain control. Assemblages of native aquatic plants have returned to the shores of public waterways and the fisheries remain vibrant.

Hydrilla, introduced to Florida in the early 1950s, has had ecological and economic impacts parallel to water hyacinth.

In its interior hydrilla mats contain little oxygen, raise surface water temperature, cause wide fluctuations in pH, and reduce the rate of decomposition by reducing oxygen and wave action. Once established, hydrilla eradication is difficult if not impossible. Like with water hyacinth, no control is initiated without considering impacts on fish and wildlife and their habitat. Most hydrilla control in Florida public waters is conducted using herbicides, as none of the five insect species or grass carp have been efficient to control this rapidly spreading plant. Mechanical control has proved too slow and costly. Using fluridone herbicide, 10 100 ha of surface-matted hydrilla on 10 930 ha Lake Istokpoga were controlled within two years, for US$ 1.8 million, which is US$ 90 ha^{-1}, as against the estimate of US$ 1 250 needed if mechanical harvesters (200 of them) would work for two years to achieve the same result. Mechanical harvesters also remove substantial numbers of fish, reptiles, amphibians, and invertebrates in addition to plants. Schardt (1999) also points out that chemical control using fluridone, and occasionally endothall herbicides, retains viable fisheries and habitat. Sport fisheries have rebounded in lakes in which repeated large-scale herbicide treatments are being applied and native aquatic plants return.

For the impact of herbicide application on aquatic plants, and changes in the aquatic environment resulting from the subsequent plant decay.

Physical (mechanical) Control of Aquatic Macrophytes and its Impact on Fish

Removal of aquatic plants is done for various reasons, such as to clear the passage for ships, to speed up the flow of water through canals, or to prevent the blockage of intakes to pumps or to penstocks of hydropower stations. Mostly the physical removal of aquatic plants is temporary, as this is followed by a regrowth, or by invasion of plants from other areas of the water body. In temperate countries, such as in the UK, Russia and the USA, annual weed cutting programmes constitute a regular maintenance cost. Excessive growths of aquatic macrophytes are common in the nutrient-rich and unshaded lowland streams and rivers of southern England. Dawson *et al.* (1991) studied what amount of aquatic invertebrates is removed with the cut weed and what consequences this may have on fish. In the River Frome, where weed cutting occurs twice a year, the biomass of invertebrates removed (largely chironomid and simuliid larvae) amounted to 2–3 t in fresh weight. In addition to destroying part of the invertebrate food supply available to fish, the weed cutting also renders many organisms vulnerable to fish predation. Dawson *et al.* (op.cit.) were unable to estimate the net effect of the reduction in food supply for fish as it would depend, among others, on the timing of weed cutting and invertebrate life histories. The authors found very few fish amongst weed that had been removed and this was in marked contrast to the use of floating harvesters in some lakes.

Mechanical harvesting of the submersed weed *Hydrilla verticillata* in Orange Lake, Florida, entangled fish in the cut vegetation resulting in their disposal with the weeds on shore. Fish collected from the harvested weeds showed a density of 205 000 ± 35 000 fish ha^{-1}, and standing crop of 460 ± 30 kg ha^{-1} (Haller *et al.*, 1980); the estimated loss of fish to mechanical harvesting represented 32% of fish numbers and 18% of fish biomass. Fish most susceptible to mechanical removal with hydrilla were juvenile sportfish and smaller species. The monetary replacement value of the fish lost was estimated at over US$ 6 000 ha^{-1}. Mechanical harvesting of *Myriophyllum spicatum* in Saratoga Lake, in the state of New York, was less harmful to fish stocks.

The harvesting operation removed primarily bluegill sunfish (*Lepomis macrochirus*) and to a lesser degree, largemouth bass (*Micropterus salmoides*), yellow perch (*Perca flavescens*) and pumpkinseed sunfish (*Lepomis gibbosus*). In total, the harvesting of *Myriophyllum* removed about 3% of the total standing crop of juvenile fish (Mikol, 1985). In another lake in the USA, also overgrown with *Myriophyllum spicatum*, the use of a mechanical plant harvester, able to cut near the sediment surface, also had only a small incidental impact on fish, removing 36 fish ha^{-1}, primarily small bluegill (*Lepomis macrochirus*) of less than 30 mm in length (Unmuth *et al.*, 1998). But Nichols (1991) noted that the impact of harvesting of aquatic macrophytes is likely to vary tremendously between lakes, due to the differences in aquatic macrophytes, their densities, and different fish stocks. Well conducted experiments to quantify the impact of harvesting on fish are needed to understand the benefits and detriments of harvesting to fish populations.

Olson *et al.* (1998) studied the impact of mechanical harvesting of aquatic macrophytes on fish in four Minnesota lakes. Based on the results they concluded that changing the strategy of harvesting from clear-cutting the top metre of vegetation to selectively cutting deep channels throughout the lake may simultaneously improve the fishery and recreational value of a lake.

Lowering of the level of river water and speeding up water flow are secondary impacts of weed cutting on fish. The former may result in drying of attached eggs on the now exposed substrate, the latter making the flow rate too fast for the fry of some fish. The loss of invertebrates resulting from regular weed cutting in the River Frome was estimated by Dawson *et al.* (1991) to result in a loss of ca. 18% of fish production. The removal of their preferred feeding position will impact some fish species more than the other. Trout *Salmo trutta*, a mid-water feeder, will be more affected that the benthic feeder bullhead (*Cottus gobio*). A change in the traditional weed cutting to another method, which would leave, e.g. marginal strips or patches of weeds, could be more beneficial for fish and probably find support of recreational fishermen.

Mortensen (1977) found evidence that weed-cutting had detrimental effects on trout fry (*Salmo trutta*) in small Danish streams. Ostergaard (1999) reported that total cutting of aquatic macrophytes caused a decline in the trout populations, while a limited flow-channel cutting had a negligible effect in a small stream, and a positive effect in a large stream, in the latter case leading to a rise in the population of small (0+) fish. Total weed cutting in the large stream resulted in a decline in the population of large fish. In small streams total weed cutting impacted both the small and the large fish. A study on the Perry River (Swales, 1982), a tributary of the Severn River in England, found that mechanical removal by cutting of substantial part of aquatic plant stands disrupted fish feeding, reproduction and normal behaviour patterns. Cutting also removed considerable numbers of aquatic invertebrates on which some fish feed. Aquatic plants in the Perry River were directly utilised as food source by dace, and were an important spawning substrate for the phytophilous roach, perch and pike. Bankside vegetation contributes terrestrial insects when they fall into water. From an environmental point of view, 25% aquatic plant cover in English lowland rivers provides for conservation of aquatic invertebrates, fish and the plants themselves (Haslam, 1978).

In Ireland, a land-based 'mowing bucket', applied in the Grand and Royal Canals, cut the aquatic vegetation to canal bed level, thereby removing any substrate for colonisation by aquatic invertebrates. Nevertheless, macroinvertebrates increased rapidly following treatment and no adverse effect on dependent fish resulted (Monahan and Caffrey, 1996). The policy of removing obstructive vegetation from a central navigation channel, while preserving weeded marginal fringes, minimizes the impact of weed control operations on the macroinvertebrate fauna, and hence on the fish which depend on this food.

Results from weed cutting along one bank of a stream in Brittany (France) showed a negative impact on fish. One month after the plant removal (largely *Ranunculus penicillatus*), electrofishing surveys showed greater fish biomass and higher densities in vegetated areas. The density of Atlantic salmon parr (*Salmo salar*) was 30 ind. 100 m^{-2}, as compared to only 15 ind. 100 m^{-2} in habitats without macrophytes. Removal of macrophytes has thus decreased the carrying capacity of the stream for age-0 salmon parr (Roussel *et al.*, 1998).

In British canals the level of aquatic plant growth is closely related to boat trafic intensity. Where boat trafiic is heavy few plants are present and where little or no traffic is recorded an abundant flora prevails (Caffrey, 1993).

Water level fluctuation is widely used for aquatic plant management. Water can be raised or lowered, but drawdown is the most feasible and most often used fluctuation method (Nichols, 1991). Aquatic fauna may exhibit a dramatic change in species composition and biomass following drawdown and reflooding. While there may be an increase in some benthic macroinvertebrates, such as oligochaetes and chironomids, other macroinvertebrates may be adversely affected because of drying and freezing of exposed sediment. The impact on fish resulting from the changes in the benthic macroinvertebrates is poorly known. However, water level drawdown has been successfully used to increase fish standing crop, shift in species composition and to enhance spawning of some fish, such as bass (*Micropterus* spp) and crappie (*Pomoxis* spp) (Cooke *et al.*, 1988).

Serafy *et al.* (1994), who studied the impact of the mechanical removal of hydrilla in the Potomac River, Maryland (USA) concluded that mechanical harvesting, when macrophyte beds are in short supply, may not be prudent as it may have a negative impact on fish species diversity and fish stocks.

AQUATIC MACROPHYTES AS A HABITAT OF VECTORS AND HOSTS OF TROPICAL DISEASES, AND BIOLOGICAL CONTROL USING FISH

This chapter is necessarily a very brief account on the use of fish for biological control of vectors and hosts of tropical diseases. For more information the reader should consult other sources.

Use of fish in mosquito control has been well known for more than 100 years and there are numerous reports of the effectiveness of various species, and some reports on limitations. The most widely know are *Gambusia affinis* (the "mosquito fish"), and *Poecilia reticulata* (the "guppy"). A great number of other fish species, including the young of various sport or food fish, prey on mosquito larvae. For example, for the countries of the Eastern Mediterranean Region, including Somalia and Sudan, and extending eastward to Pakistan, WHO (1981) lists 34 species. Of these, eight species are used directly, and 25 species are promising in mosquito control. Those used directly are: *Tilapia zillii* (Somalia), *Oreochromis mossambicus* (Pakistan), *Carassius auratus* (Iran), *Aphanius dispar* (Oman, Somalia), *Nothobranchius palmquisti* and *N. guentheri* (Somalia), *Gambusia affinis* (Jordan, Lebanon, Syria, Iran (Islamic Republic of), Iraq, Afghanistan, Sudan), *Poecilia reticulata* (Pakistan). It is recognised that the use of larvivorous fish for vector control is a simple, inexpensive and reasonably effective measure, and should be considered as a component of integrated strategies. The use of native fish should be given preference to avoid possible undesirable implications of introduction of new fish species. It was recommended to use more than one biological control agent, as, for example, a larvivorous fish together with a phytophagous fish, or two species of larvivorous fish, such as *Oryzias latipes* and *Pseudorasbora parva*, which are surface and middle-level feeders respectively (WHO, 1982). Predation on mosquito larvae and on molluscs by fish may be less efficient in dense underwater meadows of submersed plants, and in such situations a joint use of herbivorous and larvivorous and/or molluscivorous fish has been recommended. In Sudan, the introduced grass carp was reported to ingest vast numbers of *Biomphalaria* snails along with *Potamogeton*, on which it feeds (WHO, 1981). In subtropical countries, the use of grass carp in irrigation and drainage canals, and in fish ponds, is known to reduce maintenance costs by clearing the weeds, and to reduce mosquito breeding through increased flow of water or predation.

As an example from Africa, in Volta Lake in Ghana, West Africa, *Pistia* was found to harbour many mosquito larvae, among which *Aedomyia africana* dominated the fauna (49.5% of the total), followed by *Ficalbia splendens* (37.4%) and *Mansonia africana* (8.7%). *Anopheles funestus*, a malaria vector, and *Aedomyia africana*, a vector of yellow fever, encephalitis and filariasis, were found commonly associated with *Pistia* in Volta Lake (Obeng, 1969). The aquatic vegetation in Volta Lake has greatly enhanced the conditions for mosquito vectors of malaria and filariasis, and for snail hosts of schistosomiasis. Numerous laboratory and field experiments have shown that aquatic plant cover provides anopheline larvae with refuge from fish predation. In California, Orr and Resh (1987) have shown that *Potamogeton pectinatus* reduced the effectiveness of the larvivorous fish *Gambusia affinis* in controlling anopheline mosquito larvae, as the plant provided cover as a refuge from predation.

Water hyacinth in Malaysia, where this plant is widespread, has been observed to harbour insect vectors of filariasis, e.g. *Mansonia uniformis, M. indiana* and *M. annulifera* (Bakar *et al.*, 1984). An estimated 200 million people are infected worldwide by human schistosomes, worm parasites which are transmitted by various species of freshwater snails, which are frequently associated with aquatic macrophytes. The extent of infestation of human population can be seen on the example of Volta Lake, a man-made reservoir in West Africa.

In 1965, in the one-year-old Volta Lake, *Pistia stratiotes* supported populations of the gastropod *Bulinus forskali*, which is not a host of

Schistosoma haematobium (Petr, 1968). However, in the following year, *Bulinus truncatus rohlfsi* was found both on *Pistia* and *Ceratophyllum*, while *B. forskali* was diminishing in numbers. While pre-impoundment studies showed the prevalence of infection in villages and townships situated on the Volta River ranging between 1% and 3%, with one locality on a tributary to the Volta River having 8%, in the Volta delta the disease was endemic, affecting in many villages over 80% or even 90% of the school children (Paperna, 1970). By 1968 the infection in some human shore communities on Volta Lake was as high as 99%. The disease was apparently introduced to the lake by the Ewe fishermen from the lower Volta, who became attracted to the rich fishing grounds in the forming reservoir and settled in small and large camps along the shores, occasionally close to the old villages or the areas of new settlements (Paperna, 1969).

In the 1970s *Ceratophyllum* was the major habitat for *B. t. rohlfsi* (Odei, 1979), with the snails also found on many other substrata, including the lake bottom, stones, twigs and particularly palm leaves, used to make fish traps. The snails are dispersed by a number of human activities, particularly fishing. Nets often become entangled and fouled with weeds, and cleaning of nets near villages brings in the snails. Odei (1979) observed a rapid snail build-up on reflooded drawdown areas of the lake, but where there was no offshore *Ceratophyllum*, there was usually a marked decrease during the dry season, as already observed by Paperna (1969, 1970). Nothing is known about the efficiency of the local fish in controlling the snails. Flow regulation downstream of Volta Lake has caused an explosive growth of *Potamogeton octandrus* and *Vallisneria aethiopica*, and the associated *B. t. rohlfsi*. This has naturally maintained the high level of schistosomiasis, which was widespread already prior to regulation, as mentioned above.

It is not known what proportion of fish food the mosquitoe larvae and bilharzia snail vectors represented in Volta Lake. Predation on mosquito larvae and on molluscs by fish may be less efficient in dense underwater meadows of submersed plants, and in such situations a joint use of herbivorous and larvivorous and/or molluscivorous fish has been recommended. In Sudan, the introduced grass carp was reported to ingest vast numbers of *Biomphalaria* snails along with *Potamogeton*, on which it feeds (WHO, 1981). In subtropical countries, the use of grass carp in irrigation and drainage canals, and in fish ponds, is known to reduce maintenance costs by clearing the weeds and reducing mosquito breeding through increased flow of water or predation. (See also below).

Biological control of snails using fish has been practised on a limited scale in fish ponds, small dams, irrigation canals, and in combination with rice-fish culture. In Africa, several species of fish are malacophagous (*Serranochromis* sp., *Astatoreochromis alluaudi, Sarotherodon melanotheron, Clarias* sp.) but rigorous observations on their efficiency in field situations are still lacking (McCullough, 1981). He recommended that research be undertaken on the efficacy of snail-eating fish, either alone or in combination with herbivorous fish species. But even in 1995, only a few well documented field trials existed, which would report varying degrees of success with species such as black carp (*Mylopharynodon piceus*), shellcracker sunfish (*Lepomis microlophus*), and African cichlids *Astatoreochromis alluaudi* and *Haplochromis mellandi* (Slootweg, 1995). *M. piceus* was effective in controlling nuisance snails in reservoirs in Israel, *H. mellandi*, a good table fish, was successful in controlling *Bulinus* and *Biomphalaria* in fish ponds, irrigation canals and ricefields in Zaire (but Slootweg comments that this information dates from 1956), and *L. microlophus* successfully controlled snails in Puerto Rico. Slootweg also gives examples of the impact of introduced *Astatoreochromis alluaudi* on snails in small water bodies in various parts of Africa. In Kenya, where the fish was introduced in earthen dams in 1955, it initially reduced particularly *Bimphalaria pfeifferi*, but less *Bulinus* spp. *B. pfeifferi* formed the principal diet of *A. alluaudi* (Slootweg *et al.*, undated manuscript). When the dams were revisited in 1986/87, the fish were still present in 5 dams, but snails were found in abundance. In a field trial in Cameroon, which started in 1988, *A. alluaudi* was stocked together with catfish *Clarias gariepinus* and tilapia *Oreochromis niloticus*. Neither catfish, a known snail eater, nor *A. alluaudi* had any influence on resident snail populations in ponds. The conclusion of the study was that fish culture under good nutritional regimes enhances growth and reproduction of snails. Because of a lack of competition for food the so-called moluscivorous fish prefer to eat "easier" food items, readily available in fish ponds. Field trials in drainage canals and in a rice-fish culture experiment gave additional evidence that the fish were not capable of controlling snails. Slootweg (1995) gives three reasons for failure of *Astatoreochromis alluaudi* as: foraging behaviour and prey choice, plasticity of pharyngeal teeth, and snail ecology. He believes that other species of snail-eating cichlids will also fail to control molluscs for the same reasons.

Laboratory experiments showed that *Oreochromis mossambicus* and especially *Tilapia rendalli* may be promising biological control agents of schistosomiasis by acting as predators of eggs and young *Biomphalaria glabrata* less than 10 mm in diameter (Graber *et al.*, 1981). Another approach on how to control pulmonate gastropod hosts of schistosomiasis has been suggested by Daldorph and Thomas (1991). Eutrophicated water bodies are characterized by a retreat of macrophytes and their replacement by phytoplankton. While eutrophication is generally considered to be detrimental to conservation and water management in the developed world, it could possibly be of benefit in tropical countries: a collapse of the macrophyte community due to eutrophication could cause a decline in pulmonate snail population. How this would work in tropical waters is not quite clear at present. There is anecdotal evidence that *Bulinus truncatus*, a host of schistosomiasis in Egypt and the Sudan, is adversely affected by increased levels of eutrophication, but *Biomphalaria alexandrina* in the same water bodies appears to be increasing and hence benefiting from the ecological changes associated with eutrophication (Daldorph and Thomas, 1991).

Slootweg (1995) noted that if fish were to be used in snail control, it should be limited to permanent habitats and in combination with other control measures. The role of fish must be seen as part of an integrated approach where habitat alterations and appropriate water management can reduce snail breeding and refuge sites, and where natural or introduced competitors and predators put further pressure on snail population. Obeng (1978) suggested control of snail hosts of *Schistosoma* spp by using herbivorous fish such as grass and common carp to control aquatic plants and hence removing the substrate with which the snails are usually associated. Clearing of aquatic weeds using grass carp in irrigation canals in Egypt had a significant reducing effect on snail populations. Clearing of aquatic weeds reduces the amount of food and also exposes snails to predators that might be naturally present (Slootweg, 1995). A combined use of malacophagous and herbivorous fish is one of the promissing approaches, as applied already in some water bodies for the control of mosquitos. In the irrigation system of Mansuriya District, Egypt, Van Schayck (1985) noticed that grass carp had a positive effect on reducing the number of *Bulinus truncatus* and *Biomphalaria alexandrina*. Slootweg (1995) pointed out that future research activities should concentrate on integrated research, rather than hoping to find a snail eating fish that will fully eradicate intermediate hosts of schistosomiasis and other trematode parasites.

CONCLUSIONS

Research on fish and fish stocks in freshwater and brackish water habitats is often carried out in isolation. Fish are closely dependent on their total environment, the balanced condition in which can be distorted not only by human activities, but by excesive pressure evolved by introduced exotic organisms. Aquatic macrophytes are one of the major links in the interrelated environment, and some authors believe they are the central link which determines its stability. They have a central position in the network of ecological relations between nutrients, plankton and macroinvertebrates, and in the temperate zone of Europe they determine the carrying capacity for the population of some fish such as pike, most cyprinids and eel (De Nie, 1987). Much of the literature focuses on individual species or restricted groups, and there is a shortage of studies of whole communities, containing more than two trophic levels (Jones *et al.*, 1998).

Aquatic macrophyte-eating fish species are common in Europe and Asia, Africa and Latin America, but there are no fish native to North America for which aquatic macrophytes are an important dietary component (Lodge *et al.*, 1998). But aquatic macrophytes are also subject to herbivory by invertebrates, especially insects and snails, by some mammals, turtles and birds.

Literature which deals with aquatic macrophyte-fish interactions often concludes that intermediate plant densities enhance fish diversity, feeding, growth and reproduction of fish. However, "intermediate density" is a subjective term and recent evidence suggests that both spatial distribution and relative abundance of plants must be considered in plant management strategies. Management decisions are usually based on macroscale, which implies either an entire water body or a water body divided into zones (e.g. littoral, cove). Aquatic plants are mostly assessed on the basis of their area cover and expressed as hectares, percent coverage, or biomass per hectare. Fish are expressed as standing crop, density, catch per unit effort, or percent abundance relative to plant coverage. Macroscale information rarely allows to fully understand what is actually happening within the macrophyte habitat, where behavioural characteristics of fish often determine the complex relationships of temporal and spatial character during the ontogenesis of the individual fish species. Microscale studies that quantify fish behavioural responses such as habitat preference, foraging efficiency, predator avoidance, and social attraction in vegetated areas are required to clarify the role of aquatic plants as fish habitats (Dibble *et al.*, 1996).

The relationship between plant and fish abundance is an important management consideration when plant control strategies are developed, because it provides a basis to evaluate potential effects of the control programme on the integrity of the fish community structure (Killgore *et al.*, 1989). Artificial structuring of plant beds, as for example by clearing a series of deep channels (Olson *et al.*, 1998) or boat lanes, suggests that plant beds could be engineered for increased fishery value. Studies of fish-plant interactions will assist in a better understanding of causal relationships between the structure of plant beds and a specific response from the fish community and verify protocols for enhancement of fish habitats coincident with control of plants (Hoover *et al.*, 1993).

Aquatic macrophytes can contribute to an increase in fish abundance, particularly in areas once devoid of any substantial amount of cover. In the USA Borawa *et al.* (1979) found that fish densities increased from approximately 1 000 to more than 15 000 fish ha^{-1} after the invasive plant *Myriophyllum spicatum* became established in Currituck Sound. Killgore *et al.* (1989), in their study in the Potomac River, estimated fish density from 17 000 to 98 000 fish ha^{-1} in areas with plants and the CPUE was two to seven times higher in areas with aquatic plants than areas without plants. In Orange Lake, Florida, Haller *et al.* (1980)

and Shireman *et al.* (1981) found a density from 13 000 to 205 000 fish ha^{-1} in areas with submersed plants.

An intermediate density of aquatic macrophytes is still considered as preferable, from which both fish and crayfish benefit most. But Hoyer and Canfield (1993), who investigated 60 Florida lakes for the relationships between aquatic macrophytes and fish stocks, cautioned that only potential for depressed fish stocks exists at both high and low levels of aquatic macrophytes.

Persson and Crowder (1998) produced a summary of potential mechanisms by which fish may affect macrophyte abundance, littoral-pelagic coupling, and overall lake dynamics:

1. Routes for effects on submersed vegetation (habitat structure) by fish feeding activities in macrophyte habitats feeding on macrophytes feeding on epiphytes feeding on macroinvertebrates feeding-induced uprooting of plants
2. Routes for effects on submersed vegetation (habitat structure) by feeding activities in open water zooplankton predation induced changes in phytoplankton biomass (transparency) sediment feeding induced fluxes of nutrients to open water affecting phytoplankton biomass (transparency)
3. Other fish induced coupling of littoral and pelagic habitats transport of nutrients/organic matter to pelagial due to littoral feeding and pelagic excretion/egestion transport of nutrients/organic matter to littoral due to pelagic feeding at night by daytime refuging juvenile fish recruitment of juvenile fish to the vegetation habitat from pelagic larval stages over ontogeny recruitment of adult fish to the pelagic habitat from the vegetation habitat over ontogeny.

Researchers are now focusing more on the whole submersed macrophyte food web. The trophic cascade, which has been demonstrated for relatively simple food chains, sometime fails in more complicated webs. Crowder *et al.* (1998) highlighted the need for linking the littoral and pelagic habitats both in terms of prey consumption and translocation of nutrients, to include for example the transfer of epiphyte productivity to pelagic habitat. Macrophytes, as a substrate for epiphyton, and fish, are the major factors in such a simplified system, which shows the importance of a vegetated littoral for the whole ecosystem of a water body.

The need for biocontrol of aquatic macrophytes/weeds still remains high on the list of priorities, not only for fishery managers, but also for maintaining good quantity and quality water for drinking and industrial uses, for engineers of hydroelectric dams, for river and lake transport etc. The need for aquatic weed control can be seen from the situation in India where in the mid-1970s about 320 000 ha of water surface had to be cleared of weeds annually for fish culture, and in eastern states up to 70% of water area was infested by weeds (Chaudhuri *et al.*, 1977). Since then, the situation has probably further deteriorated. Submersed and floating aquatic macrophytes are also a habitat of vectors and hosts of parasitic diseases, but the notion that fish may become an efficient control agent has been revised as it is becoming evident that only integrated control appears to be the solution to the problem.

Fisheries represent only one of the users of the aquatic resource. Wiley *et al.* (1984) stressed that management strategies for aquatic macrophytes should reflect both the ecological realities of aquatic systems and the practical uses for which water is intended. If recreational or commercial fish production is an important goal, fisheries management and aquatic plant management must be integrated in a compatible way. Barko *et al.* (1986) pointed out that anagement procedures directed toward aquatic macrophyte eradication are generally ill-advised, not only on ecological grounds, but also because many aquatic systems serve multiple purposes. Major environmental changes are generally related to the amount of vegetation controlled rather than to the control method. Control of aquatic weeds should, therefore, be a major consideration in selecting control methods, especially when monetary resources are limited (Sutton, 1985). Wide information dissemination, including good case studies, and training using a holistic and integrated approach to ecosystem management are needed for faster progress in this field. The start has been already made, as seen from publications such as Jeppesen *et al.* (1998a) and Gulati *et al.* (1990).

Much of the research on interrelationships is done in developed countries. There exist centres of excellence, such as in the USA (Florida),

Denmark, Germany and Brasil, where most interesting work has been done in the last 10–30 years. Transfer of knowledge from industrialized (often temperate) to less industrialized (often tropical) countries is gaining speed, but not all the knowledge gained in and on waters in a temperate climate is relevant or applicable. It is mostly the latter countries where problems with invasive aquatic macrophytes, such as water hyacinth and salvinia, are most urgent, as such plants interfere with water supply and inland fisheries, both of which are vitally important in tropical countries. For some time to come, industrialized countries may remain the vanguard of indepth, sound, and applicable research and of devising the best ways of aquatic plant management for water bodies of the temperate and tropical climate. But it would be on the countries situated in the tropics and subtropics to critically validate the knowledge already available and test it in local conditions, but also to undertake both field and experimental studies on the function of aquatic plants in determining the level of fish productivity in vegetated water bodies.

Bibliography

A.D. Dholakia: Fisheries and Aquatic Resources of India, Daya, Delhi, 2004.

Arnold C. Long: Deep Sea Demersal Fish and Fisheries, Cyber Tech Pub, Delhi, 2009.

B.R. Selvamani and R.K. Mahadevan: Marine Capture Fisheries, Campus Books, Delhi, 2008.

C.B.L. Srivastava: A Text Book of Fishery Science and Indian Fisheries, Kitab Mahal, Delhi, 1999.

K.P. Biswas: Economics in Commercial Fisheries, Daya, New Delhi, 2006.

Omprakash Sharma: Handbook of Fisheries and Aquaculture, Agrotech Publishing Academy, New Delhi, 2009.

Rajendra Kumar: Biotechnology and Genetics in Fisheries and Acquaculture, Arise Pub, Delhi, 2010.

Ramakrishnan Korakandy: Fisheries Development in India : The Political Economy of Unsustainable Development, Kalpaz Pub, Delhi, 2008.

S P Malhotra and V R P Sinha: Indian Fisheries and Aquaculture in a Globalizing Economy, Narendra Pub House, Delhi, 2007.

S. Nath and S.C. Dey: The Ecology and Fisheries of Barak River System of India, Narendra Pub, Delhi, 2008.

Tomi Petr: Fisheries in Irrigation Systems of Arid Asia, Daya, New Delhi, 2007.

Umesh C Goswami: Natural and Anthropogenic Hazards on Fish and Fisheries, Narendra Pub House, Delhi, 2007.

V B Sakhare: Reservoir Fisheries and Limnology, Narendra Pub, Delhi, 2007.

V R P Sinha: Fisheries Research Planning and Management in Developing Countries, Narendra, Delhi, 2005.

Index

□□□